A CHRISTMAS GIFT FROM MY
VERY DEAREST FRIEND
MARGARET SHARP

25th DEC. 2008

HIGHLAND YEAR

HIGHLAND YEAR

L. MacNally

ILLUSTRATED WITH 83 PHOTOGRAPHS
BY THE AUTHOR

PHOENIX HOUSE • LONDON

For Margaret
who has put up with the peculiarities
of a stalker-naturalist-photographer for many years
and for Bill Beaton
who fostered in me a love of the outdoors

Made in Great Britain at the Aldine Press, Letchworth, Herts
for J. M. DENT & SONS LTD, Aldine House, Bedford Street, London
A Phoenix House publication
First published 1968
Reprinted 1968

SBN: 460 07727 9

Contents

ACKNOWLEDGMENTS

I am grateful to the many friends without whose active encouragement this would not have been written, in particular Andrée and Louis Petyt, Edgar Barclay, Philip Brown and Henry Tegner.
I am grateful also to the editor of the *Field*, *Shooting Times* and the *Scots Magazine* for permission to include material previously used in their pages.

Illustrations

AUTUMN INTO WINTER

between pages 8 and 9

WINTER INTO SPRING

between pages 40 and 41

SPRING INTO SUMMER

between pages 72 and 73

SUMMER INTO AUTUMN

between pages 104 and 105

Autumn into Winter

1

November

LIFE IN THE HILLS

LIFE in an isolated house, tucked in below a western fringe of the Monadhliaths and almost three miles from the nearest village, has an attraction for me which far outweighs the few disadvantages, such as the occasional day in winter when snow may block the road so that neither mail nor grocery van can reach us.

The view for one thing! A few steps from my door and a magnificent vista opens out down the Great Glen with the fiord-like tongue of Loch Ness reaching out, cradled in hills, to the horizon. With each season the view changes in enchantment, from the freshness of spring, through the lushness of summer, to breathtaking autumnal splendour, reaching its climax in the often savage grandeur of winter's snow-streaked hills and steel-grey loch.

Loch Ness has the distinction, perhaps not altogether enviable, of being known throughout Britain as the home of 'the monster'. To me, the fact that its banks afford shelter to a rich variety of wild life is much more of an attraction than any monster. The existence of 'something' in Loch Ness was held as a matter of course by the old folk of the district, and one veteran ex-stalker told me that it was never referred to as 'the monster' then but simply as 'the big beast'! The fact that it was seen on occasion was accepted without fuss or publicity until the newspapers got hold of it and it became a cause for country-wide speculation. I myself, though I was born near Loch Ness and have spent most of my life overlooking it, have never seen the monster. But my wife, not a native of the district, has—and this only a year after she came to live here, a sighting which she has never publicized, and indeed has told to no one but myself. She saw it on a clear sunny June afternoon as she walked down the road from our house with Loch Ness, still as a millpond, before her. An object, as she later described it to me, like the dark-coloured head and neck of a giraffe broke the calm, still surface of the loch and proceeded at speed across it, leaving a V-shaped ripple behind it. As suddenly as it had appeared so did it submerge, and within moments the loch was like a mirror again. I envy my wife this sighting, and though I have never seen the monster myself I am convinced that there is something strange in Loch Ness. Many reputable people of my acquaintance have seen 'something'; too many for me to presume to deny its existence.

Turning from the view down Loch Ness to face eastwards, a steep hill face rises opposite my windows from the river hidden in a wooded gorge below. Tree-clad in birch and hazel, interspersed with giant ash trees on its lower slopes, its top frets the sky in rock-strewn wildness. On the stage of this face, from bedroom window or from fireside chair, I have watched all three species of deer which exist here, red deer, roe deer, or Japanese sika deer, more especially in winter, dark against the snow as they forage for food. Fox, badger and wildcat have appeared fleetingly, and the golden eagle has hunted along its face while the 'mewing', soaring buzzard is commonplace. From it, in appropriate season, the skirl of vixen or triple bark of dog fox, the chilling savagery of wildcat caterwauling, the gruff bark of roe, rutting roar of red deer stag, or thrice repeated ear-splitting rutting whistle of sika stag, have reached my ears. Not for me loneliness in such a situation.

Winter is one of my favourite seasons and is perhaps the best of the four seasons for seeing the wild life around us. The heaviness of summer growth has gone and bad weather may bring the wild creatures near to our door. So many people come north in the summer and go away disappointed at not having seen deer or eagle, essence of the Highlands to most of the tourists, which in high summer can seem almost non-existent. Both eagle and red deer are then away on the high tops, remote from the tourist roads, and the lower hills, which seem forbiddingly high to most Southrons, are empty. How much more would they see if they came in winter; but to see it they would have to brave the vagaries of Highland weather at that season, a prospect lacking in attraction to most people.

Winter weather in the Highlands, mind you, is never to be taken lightly, as I personally found out and luckily survived to tell the tale. I had been out, on a day of deep, soft snow, stalking hinds. The glen bottom, white and featureless, was empty of deer, and I had to climb the hill face to where a wide flat stretched behind it, a flat seamed and channelled by peat hags. Reaching the top, an unbroken prospect of white stretched before me, every landmark changed and obliterated by the depth of snow, while a spindrift of fine, gritty snow was being wind-blown across it. It was bitterly cold in that wind and I got going across that flat as fast as the clinging depth of snow and stinging head wind would allow. Without warning, about the middle of the flat, I floundered to my knees in seemingly bottomless snow. My rifle was slung on my back or I must have lost it in my sudden fall. Not yet realizing the seriousness of my situation I at once tried to get to my feet, only to find that as soon as I put my weight on one foot I started to sink and could find no purchase or bottom in the snow. I tried to edge myself forward on my knees and again began to sink. Only where I had packed the snow hard below my knees in my fall could I remain without sinking. My next attempt, at crawling out on hands and knees, was wellnigh disastrous. As soon as I put weight on my outstretched widespread hands I sank quickly forward until my face was pressed hard into the white, cold softness, nose and ears full of snow. It was with the utmost difficulty and in near panic that I somehow levered myself back to my former position, on my knees. I had realized by this time where I had landed. I had stumbled in the featurelessness of that white expanse into the middle of a large, deep peat hag which was drifted full of soft snow. I was in a drift of about seven feet of snow with an unknown depth of semi-liquid peat below that. Fear began to oust reason; was I to freeze thus on my knees or struggle and smother in a welter of snow and peat?

No matter how I manœuvred I simply could not, absurd as it may sound, regain my feet. My fall had taken me well out into the drift and there it seemed I was stuck. There was no one out on the winter hills but myself, no shepherd even, as the sheep were lower down. It would be dark before any anxiety was felt, and even then my exact whereabouts would be unknown. Rescue would probably arrive too late; in that wildness of wind-blown snow my tracks would have been covered long ago. Almost in despair I remembered reading of how, in quicksand, one should lie flat and present the largest area of one's body to its softness. Reluctant though I was to try it, remembering all too vividly the suffocating feeling as my face sank in the snow in my attempt at crawling, something had to be tried, and quickly. I unslung my rifle and gripped it at the muzzle with one hand and at the butt with the other. Then I gradually eased forward and extended rifle and arms ahead of me. I pulled and edged myself bodily forward, and to my joy made headway without sinking too much. My progress was painfully slow and to any onlooker must have looked ludicrous, a human frog spreadeagled, breast-stroking his way across the snow.

At length I judged I must be clear of the hag and tentatively essayed a leg below me. What glorious relief to find, after an initial sinking, solid ground below me, and with what care I made for the nearest hard ridge, avoiding like poison any suggestion of drain or hag. So buoyant is the human spirit that I spied, stalked and bagged two hinds from a herd on my way home. In retrospect, the whole episode has an element of the ridiculous in it, a stalker stuck in snow on hills he knew intimately. Only a hair's breadth, however, divided the ridiculous from the finality of another winter tragedy on the Scottish hills.

It was while hind-stalking too that I witnessed the fear which the golden eagle can inspire in adult red deer. Walking along the path midway up one side of a tree-clad, river-bottomed glen I saw deer begin to pour over the skyline of the opposite side of the glen and run in a panic-stricken ragged bunch into the trees below. There they scattered and stood beneath the bare branches while I wondered exceedingly. And then there floated lazily and, it seemed, almost insolently, into view above that skyline first one, then another eagle. This then was what had stampeded the deer down into the shelter of the trees, knowing full well that they were safe there from the eagle's swoop. That the eagle will kill deer calves I know, but I doubt if it will often attempt seriously to attack adult deer. Nevertheless the ingrained fear is there, possibly a legacy from calfhood sufficient to cause panic on occasion, such as I had witnessed.

A close friend of ours, while out hind-stalking with my brother one November day, had an experience with an eagle which I would dearly love to have had. The hill that day was patched with snowdrifts, remains of an earlier snowfall and, while waiting alone, he saw an eagle swooping repeatedly at something on the ground. He immediately spied at it with his telescope and to his surprise saw distinctly outlined against a patch of snow a large wildcat, crouched at bay, tail fluffed out and curled up over its back, every hair standing out, awaiting the eagle's next swoop. The swoop came and the cat reared up and swiped right and left, lightning swift, at the eagle. The eagle, showing a healthy regard for the cat's 'fish-hooks', pulled out of its swoop and, winging round, essayed another swoop, to be met again by the defiance of the wildcat. While my lucky friend watched enthralled, these tactics of attack and defence were repeated some half-dozen times before the eagle withdrew and left the cat in possession of the field. Such encounters between well-matched predators (a hen eagle will weigh around 11 lb. while the average wildcat will weigh 11–14 lb.) are of course rare. I have no doubt whatever that an eagle could easily kill a wildcat if it took it completely by surprise from above and struck true with its formidable talons; yet if the cat, with those needle-sharp feline reflexes, did have enough warning to avoid the first attack, stalemate, as in the case witnessed, would almost inevitably ensue.

My own closest encounter in winter with an eagle was much tamer and less rewarding. I had had to leave a dead hind out overnight and was returning next day with a pony to collect it. Out from a tree only fifty yards away, as I neared my hind, an eagle flopped, literally flopped, and flew heavily away. The reason for its heaviness of flight was only too apparent when I reached my hind. The ground for yards around was littered with deer hair, while a lattice work of bare, picked ribs stared up at me. The eagle had dined sumptuously. An eagle will gorge itself, vulture like, on carrion when other prey is scarce. A shepherd I knew of once pursued an eagle which had gorged itself so well on a dead sheep that it could not take wing from the flat bottom of the glen. Along the glen the chase progressed, the eagle only just evading the swipes of the shepherd's stick until, when it seemed he almost had it, the eagle gained a slight rise in the ground, sufficient to enable it to get airborne. The eagle, though popularly believed to eat only what it kills, is by no means averse to carrion, and will eat a great deal of it in winter when other prey is scarce.

2

December

STALKING THE RED DEER HINDS

THE stalking of red deer hinds, the major winter occupation of the Highland deer-stalker, is not, as may be popularly supposed, simply a blood sport, a victimization of the graceful deer by unfeeling sportsmen. There are in the Highlands today virtually no predators except man able to kill adult red deer. This being so it has become a duty to take from the deer herds each year sufficient numbers, selectively and discriminately, to keep the annual increase within the bounds of the feeding available, and so ensure that the herds are able to subsist, comfortably, *in winter* on their hill grazing. The alternative is to allow numbers to increase until starvation becomes the controlling agent, while howls go up all over the Highlands about hungry deer raiding arable ground.

Apart from this very necessary control aspect of hind-stalking, the venison secured thus is a practical annual asset, a cash crop of excellent meat in a world said to be short of meat.

The Highlands without deer would be unthinkable, but to have deer, under present-day conditions, with afforestation and other interests progressively encroaching on former deer wintering ground, one must manage them, for their good as well as our own.

It has been established by Nature Conservancy research that one needs to take one-sixth (16 per cent) of each sex, excluding calves, to

Dark against a white world, the stalker sets off for the hill.

I

Red deer stags in winter with the snow level well down on the hills.

A golden eagle, a hen, takes off from her eyrie in the high tops, showing the tremendous size of her wings. A female eagle may have a wing-span of 7 ft.

A herd of red deer hinds down at the tree-line in winter snow, crossing a partly frozen burn and stringing up the farther bank.

Opposite, middle. The hill fox, a resourceful and beautiful predator, relaxed and at rest, brush curled around.

Opposite. The Scottish wildcat, *Felis sylvestris grampia,* much flatter and broader in the head than a domestic cat, and with a blunt-ended bushy tail, with well-defined black rings.

Red deer hinds, grazing in light snow, are much easier to approach when they are thus engrossed than when they are lying resting or chewing the cud.

Red deer approaching and beginning to jump a sheep fence which divides low ground from the high ground. One hind clears the fence in fine style.

Deer will lie and become as white as their surroundings while a blizzard rages about them, a pitfall for the unwary stalker as he sees an apparently empty hill become a herd of deer.
A red deer hind.

Above. A trio of red deer stags, of about four or five years old, in deep snow. *Below*. Strapping a hind on a pony in deep snow is a finger-chilling business for the stalker.

V

Winter meal for a fox, a mountain hare in full winter white.

The stoat, in its winter dress known as the ermine, was caught out, glaringly obvious when there was no snow on the ground.

With the situation reversed, the fox is conspicuous in the whiteness of the snow-clad hills as it goes about its mating affairs.

Roe twins, of about seven months, old lying together in the *bed* which their body heat has melted in the snow.

VII

A young roe deer, grazing on exposed heather tops in deep snow, showing the prominent white caudal patch, and also the tail-like anal tush which distinguishes the doe. A buck lacks this tush. Roe are our smallest indigenous deer, about 30 in. at shoulder height, compared with the 48 in. or so of a red deer stag.

A red deer calf about seven months old nibbling at lichen on a birch trunk in deep snow. A bright frosty day in January.

A Japanese sika stag (*left*) showing typical narrow antlers. The photograph was taken in late September when the dapples of the summer coat were being replaced by the dark winter coat. Sikas are between red and roe deer in size.
A red deer knobber (*right*), a male too young to have achieved antlers but with *knobs* showing where antlers will eventually burgeon.

balance the annual increase. The stalker's task in winter, then, is to assess his hind numbers and exact this 16 per cent toll. He should do this with as much discrimination as possible, culling old and weakly deer for the ultimate good of the herd. These poor-quality deer will be obvious to the trained eye, thin or undersized, red, harsh-coated beasts, lacking the polish of a winter-coated hind in good condition. There are always, under Highland conditions, a goodly number of yeld hinds in each herd, i.e. hinds which have no calves at foot in that particular season. Some of these will be young hinds which have not yet reached calf-bearing age, some will be mature hinds 'resting' after successive years of calf-bearing, and some will have lost calves through predation or other causes. These hinds will be in the very best of condition, good looking, thickset and with a lovely 'bloom' on their coats, and when best quality venison is wanted these are the beasts to take. I believe, however, that it is a bad mistake to shoot too many yeld hinds, as these hinds are the ones which will winter best in that particular season, having no calf to support, and are invariably carrying larger calves which will usually be born earlier and get a better start in life than those of the hinds which have had a calf at foot all winter. It is also a mistake to levy your toll on the deer easiest to get at; it should be spread out over all the deer herds on your particular ground.

This shooting of hinds, in the bleakness of the Highlands in winter, can be arduous and a test of stamina, particularly if the stalker is single-handed on a large estate as so many are nowadays. The actual open season for hind-shooting is from 21 October until 15 February, but I favour a shorter period, from 15 November until 31 January. One has to be out in all weathers or risk falling behind in one's numbers. To fall behind badly may mean trying to catch up by shooting a lot of hinds in a few days. This can be done sometimes, but it is the death knell of *selective* shooting and should be avoided.

You may think you know the worst that winter weather can offer, but do you?

Have you ever lain out in a chilly bed of soft, clinging snow, within shot of a herd of hinds, striving to keep rifle muzzle and sights clear of snow while trying to pick out a suitable hind in the worsening light that presages more snow? Realizing the approaching onset you try hard to pick your beast. Still unsure, the first feathery flakes drift soft and cold in your face and, in seconds, the deer are blotted out in a whirling-dervish dance of snowflakes. If you are lucky it may not last long, and when it clears, and the deer shake off their coats of snow, you may still get your shot, if you can control your shivering limbs sufficiently.

Or have you, already wet on a dismally rainy January day, crawled up a boggy drain, or across the slow-thawing, ice-bound surface of a wide peat hag, its still thick ice holding an inch of freezing water on its top, to have to subside soddenly, almost within shot, as a watchful hind suddenly ceases chewing her cud and stretches her long neck inquiringly. Long minutes of utter discomfort and nerve-tingling immobility ensue. You lie shivering and miserably cramped, not daring to move a muscle and yet longing to rise and shout abuse at the malicious beldam who holds you pinned down.

At last her suspicions are allayed, her jaws resume their rhythmical chewing and you can crawl the last few yards to the peat hag's farther bank. Another few yards on your stomach, a cautious slithering into position, rifle sliding ever so slowly over your cover, and then a fierce visibility-blotting squall forces you into immobility again. When it finally clears you are reduced to a human icicle, while the hinds, lying couched comfortably ahead of you, are seemingly immune alike to cold or wet, with, one could swear, almost a supercilious twist to their lips as they placidly chew the cud. One by one they rise to become lost in a cloud of silvery vapour as they shake the wet from their coats before lying down comfortably again. How often have I wished I also could expel my burden of wetness thus; and how often have I wondered too whose was then the better part, hunter's or hunted. One remembers too, with a wry inward smile, remarks sometimes heard, or seen in print, of the poor hinds floundering in deep snow at the mercy of the pitiless stalker. The boot is rather on the other foot; it is the stalker who most often flounders. No matter the winter conditions on the hill, the deer are always more able to manœuvre in them than is their human adversary.

To make up for the bad days there are good days too, days on which you are glad to be alive and out in the hills, days at times breathtaking in their loveliness. After a heavy overnight fall of snow, on a day of clear, windless frost, the sun, bright in a cloudless blue sky, extracts a million scintillating twinkles from the unsullied snow, and the trees in the glen are wreathed in garments of snow. Beside your path the burn's murmur is muted under its icy cover. Here and there a particularly turbulent black pool has retained its freedom, but the more sluggish shallows are congealed into sheets of intricate frosted filigree work. Delicate ice-created 'Snow Queen' fantasies, tiny castles and châteaux crown isolated boulders in some pools. Where a fallen tree top dips into the water miniature chandeliers and nodules of clear ice coat its tips. It is an ethereal, sparkling, invigorating, new, clean world with keen, frosty air stinging your nostrils at every breath. Stop a while though even on such a beautiful day, the bright sun lending it an illusion of warmth, and the chill begins to grip you. So cold can it be, out of the sunlight, that damp woollen gloves, doffed while you gralloch a hind, can be frozen stiff when you turn to put them on again, and your fingers will sometimes adhere stickily to the ice-cold metalwork of your rifle. If you are out late on such a day, and the sun has set, it is time to be heading for home, fast, for then the implacable cold really grips, and your outer clothing, wet with 'snow-crawling', freezes stiffly and rustles in time with your every step.

Conditions of that type of frost, which has left a thin crust on top of the snow, can test one's stalking abilities to the utmost. Assuming you have managed to approach unseen to about three hundred yards or so, the real test then begins. You may have concealment all the way now to your firing point, but there is another snag. Gently as you strive to put your foot down, or silently as you endeavour to crawl or slither along, a frosty, splintering crunch bespeaks your every move. You may still have some few yards yet to go to gain your firing point when the unmistakable whisper and crackle of hooves in the snow tells you that your hinds are off. In the almost tangible silence of the frosty winter air your crunching progress has betrayed you.

A blizzard of snow, whirling over the unbroken white of the winter hills, can be deceiving as to size and scale. While walking into such a blizzard, striving to pierce the veil of multitudinous white flakes, I once saw 'a herd of deer' ahead. Down I got in the deep, soft snow and cautiously crawled to a vantage point. When the blizzard cleared I was in perfect position, rifle sights lined up on some large, scattered clumps of rushes sticking up, dark, through the snow.

When, as often occurs, one has to stalk a big herd of hinds one has the watchfulness of a hundred pair of eyes to evade. Worse, the beast you want may be at the head of the herd while you are wriggling, belly flat, at its rear. Extreme care is then needed. If even a slight movement catches the eye of only one of the herd, it is sufficient to arouse her suspicions. One then lies in a petrified state, torn between hope and despair, spiced, I'm afraid, with vindictive rage toward the persistent nosiness of your discoverer; while she, with that mincing, high-stepping, walking-on-hot-coals gait of suspicions unconfirmed, with back of rigid straightness, extended neck and widespread ears, now retreats, now advances, alerting all the herd by her demeanour, and, to make matters worse, gives vent at intervals to that most uncompromising of all sounds to a stalker's ears, the harsh, coughing bark of red deer alarm.

Grazing deer, intent on feeding, are easier to approach than deer 'lying up', chewing the cud, with eyes for any strange movement; but even here incautious, hasty tactics can lead to discovery and cause all the grazing heads to shoot up, prior to a speedy withdrawal. Short cuts in deer stalking most often lead to long days.

Perhaps most provoking of all is to walk towards a snow-clad face, immediately after a heavy snow shower has ceased, and to see it suddenly erupt deer which had been lying there, still with a mantle of the recent snow on their coats. Apparent emptiness amazingly transformed into a herd of snow-covered hinds and calves, who pause only to shake their coats free of their camouflage and fleetingly survey your now shrinking, half-crouched and naked-feeling figure, dark against the white expanse, before disappearing swiftly over the ridge top. Nothing can engender such a feeling of utter chagrin, and absolute silliness, as this Houdini-like transformation of empty hill into herd of deer. It is devastating to one's conceit, and one looks long at every sheltered face thereafter.

Having shot your hind, or hinds (I usually try for two, as two hinds make a load for one pony) the relatively easy part is over. It remains to get them home, and for this a hill pony trained to carry deer is used, equipped with a deer saddle, on which you heave your hinds (an average hind weighs about 8 stones) and secure them thereon with straps. Conditions of weather can vary enormously here too: conditions of deep, drifted snow through which one has bodily to force one's way; of quick thaw with your path holding a foot of dirty, fast-liquefying snow, every step of the pony spraying you with slush; of snow so heavy that you have to turn aside and detour high above your usual path because the snow-bowed branches of its trees deny you passage; or of gale force winds which try to blow you back to your starting point. On one such day a tremendous gust came just as I had a hind poised on the deer saddle—and draped the hind back lovingly around my neck, a gesture little appreciated. Then there are days when hinds left out overnight are just snow-shrouded humps. The utter, finger-tingling coldness of these overnight chilled carcases has to be experienced to be believed, benumbing fingers so quickly that only one strap at a time can be secured before an enforced pause to try and instil life back into icy fingers. Blizzards of snow too can transform a normally black pony into a snow-plastered dirty-white one, and oneself into a fair impersonation of an Abominable Snowman.

Welcome indeed after this is the warm fire and the nectar of a hot, sweet cup of tea, before the final work of dressing out your carcases.

Despite all that weather can do, however, the hills still exert their spell, and you get satisfaction in beating the winter conditions and doing a job you feel worth while.

So often, out among the deer, one gets many rewarding glimpses. I once had a touching example of maternal solicitude shown me by a red deer hind. I was coming along above a wire sheep fence which divided the hill ground from the glen below when I spotted a ragged-looking hind feeding on the lower side of the fence. Close at hand fed her follower, i.e. her eighteen-month-old offspring, a young, dark-coated hind. As I lay spying at them so did a herd of hinds begin to appear below me slanting uphill from the glen. I shot two of the hinds from this herd and then rose to my feet to go down to them. The herd had speedily decamped after the shots, but the ragged, red-looking hind stood her ground until I was quite near to her, and as I came in view of a section of the fence hitherto unseen I understood why. Caught by a hind leg in the fence by the crossing over of the two top wires as she'd tripped in jumping it, was her six-months'-old calf. As I went forward to free the calf the mother disappeared, reluctantly, over the fence and around the ridge behind. Her calf had obviously been trapped for at least a day. The leg just above the hoof was grooved to the bone and marked by the rusty wire. Though she seemed quite lively, the leg was very stiff through cold and constrained position. I left her lying as comfortably as I could contrive in a bed of long heather and went on around the ridge and there I met the mother, head on, almost bumping into her. Anxious about her calf she was returning into the very jaws of danger. One wonders what were her feelings about it all, the inexplicable inability of her calf to follow her when first caught in the fence and her 'solving' of this problem by grazing near by. Then after a lengthy vigil the two rifle shots signifying danger, the further alarming arrival of the human and, despite all this, her decision to return to her calf. Not every hind will show this strong maternal feeling at a time of year when maternal feelings are beginning to weaken, especially if she is one of a herd. Coming into the lower ground to graze at evening, as they do regularly, the deer have to jump the sheep fences. From the upper side, with ground advantage, it may be easy, but going out again in the morning it may prove too much for some of the smaller or less determined calves. Some half-dozen may be left inside while the herd drifts out uphill. The mothers, torn between the herd instinct and their weakening maternal feelings, often linger only a little before following the herd. The calves run up and down inside the fence, seeking an opening, poking their heads through, butting it, starting back, making half-hearted attempts at jumping, then, giving up, running inside the fence again, sharp hooves wearing a track until perhaps one, then another, manages to scramble through or over. Two or three apparently quite witless youngsters may be left inside, a commonplace sight in most winters, to be reunited perhaps in the evening when the herd returns.

Coming home early one December afternoon I witnessed another little instance of maternal feeling on the part of a red deer hind. I came, in bright wintry sunshine, over a knoll to see quite close below me the hind and her calf, the calf a puny, obviously late-born one, feeding without any suspicion of my nearness. The hind, a gaunt-looking, oldish beast, appeared bright rusty-red against the dark green of the heather. I sat down cautiously, still unseen, and spied at them, to see the hind cease feeding, approach her calf and begin rhythmically to lick its neck, her long pink tongue clearly visible through my glass. As she did so her eyes were almost closed, while the calf held its nose pointed slightly upwards, stiffly, as if mesmerized. Both were very obviously deriving great pleasure from the licking process. After about two minutes the hind ceased her ministrations and stood lazily cudding, enjoying the faint warmth of the December sun, at times closing her eyes, blissfully content. As she chewed, so close was I and so good the light, I could see she was badly gap-toothed, and could see the lump of 'cud' travel up and down her long neck as she periodically renewed it.

An old 'done' hind, with a puny, ill-nourished calf, they would be of no good on the ground, but the rifle stayed, unused, on my back. Who could shoot after watching such an intimate scene?

It is quite apparent that there is enjoyment to both licker and licked in such a case. I often wonder if the licker enjoys some salty taste of the other's coat. Certainly this licking is a fairly common practice among red deer. On another occasion, as I lay watching a small herd, a young hind and a 'knobber' (i.e. a young stag which has as yet only knobs where his antlers will be) were grazing near to each other. While I watched the young hind approached the knobber and, after standing by him for a moment, began to lick him, on his neck and under his jaw. The knobber stood perfectly still while this went on, enjoying the combined wash and brush up, the hind obviously sharing this enjoyment.

A sight with a touch of comedy I saw while watching deer lying, to all appearances very cosily, in deep snow. As I watched, a big hind arose from her 'bed', a roughly oval one formed by the heat of her body in the snow, walked across to a smaller hind and smacked her deliberately with one foreleg, causing her to get up hurriedly and vacate her bed. The big hind then lay down with a satisfied air in the vacated bed. Not to be outdone, the smaller hind then crossed to where a calf was lying near by and served it 'notice to quit'. I have no doubt that if there had been something lower down the social scale than the calf it would in turn have been victimized, but in default of this the calf just went a step or two and lay down to create a fresh bed by the weight and warmth of its body.

A good example of social dominance in the herd and one which caused me some amusement as I lay, by no means cosily, in my snowy bed.

3

January

MATING TIME OF THE HILL FOX

IN A winter largely spent out of doors one sees much of interest in glimpses of wild life denied to those who perforce spend most of their winter indoors.

A double line of tracks on the snow-sprinkled, partly frozen surface of a hill burn shows where a pair of otters have journeyed along together, diving into the ice-cold black waters where the burn was open. In deep snow one may see a furrow left by the short-legged otter as it journeyed for a time overland, literally ploughing its way along; an exhausting business it must be for an aquatic animal so absolutely out of its element.

One always remembers encounters with the elusive hill fox (and elusive the fox has to be in the Highlands with every hand against it), as on the day when, leading the pony at first light on a winter's morning, I came around a corner of the path to see a large yellow 'dog' approaching me, trotting sedately along the path in the still dim light. A yellow dog out here? Of course not! A large, yellowish dog fox! I dropped to the ground at once, as realization dawned, right under the pony's nose, trying to get my rifle from out of its cover. Hard to say who was the more surprised, the pony at my strange action, the fox at the sudden apparition of pony and stalker or myself at seeing the dog-like sedateness of the approaching fox.

At all events, it was the fox that recovered first, to run unscathed up the hillside flanking the path, while I rose from my damp bed below the wondering pony's nose, a trifle chagrined but with one more memory to store.

Wild animals in fact use the hill paths surprisingly often, fox, wildcat, red deer and roe, otter occasionally, as their tracks in snow testify. On yet another winter morning it was a wildcat which met me almost face to face as I walked quietly along a narrow path.

The most amusing encounter I have had with a fox occurred while out hind-stalking, as a result of my startling a small herd of young stags. As they trotted in single file through the long heather of a sheltered hill face some way from me, the leader suddenly jumped almost straight up in the air, and out to one side, while from below his very hooves started out a very surprised fox, drowsily confused, almost rubbing its eyes one felt, rudely awakened from its snug heather bed. The fox ran behind a little ridge without noticing me, and judging that I had a chance to cut it off on the line it was taking I ran quickly forward. As I went, cautiously now, towards a commanding knoll ahead, so did an inquiring, sharp-snouted, long-whiskered visage, ears pricked, appear over its top, to give me one fleeting, quizzical look before vanishing, all sleepiness gone.

It is in January, and into February (their mating time) that foxes are at their most vocal, and this mainly in the long hours of winter darkness. From the rugged, tree-grown hill face across the river from my house I may then hear, at first light or in the gloaming, the unearthly love skirl of the vixen, or, more often, the *staccato* double bark of a questing dog fox. In the darkness of one January night I listened to a vixen skirling repeatedly, while from two different directions two suitors 'barked' in answer and drew nearer at each skirl. Dog foxes appear to outnumber vixens here, perhaps because the vixen is the more vulnerable at the annual den offensive at the cubbing time, and the hours of darkness probably cloak many a sharp-fanged struggle to decide who wins the 'red lady'. A shepherd here once told me of seeing a vixen coming through the hill escorted by a large dog fox, while close behind, too close for his own well-being, trailed a younger dog fox. Every so often the larger dog fox raced back and trounced the younger, but squeal and squirm though he did, the lure of the vixen was apparently too strong, for trail along he would, to be trounced soundly again whenever he ventured too close.

Vulpine courtship goes on night long, the fox being largely nocturnal, and I have awakened with a start in the early hours of a January morning to hear all my terriers barking, with a 'let me get at you' note, while a vixen holds concert across the river.

The betraying snow at this time sometimes shows the neat double line of tracks of a mating pair as they 'honeymoon' through the hills, with here and there a flurry in the snow where they have presumably mated.

One of the questions which crop up every so often among people interested in wild life is whether dog fox ties to vixen in mating, as with domesticated dog and bitch. I have never personally witnessed the mating of a pair of hill foxes, but a reliable colleague once told me of how he watched this occur, through his stalking glass against a background of snow, and that they definitely tied at the consummation of their mating. I have also been told of a shepherd who came suddenly into a little hollow, while out on the hill, to find himself almost on top of a pair of tied foxes, both of which he killed before they could separate. This, if true, must have been a million to one chance.

I do know as a definite fact of two stalkers, one of whom lay and watched through his stalking glass his colleague attempt to stalk, again in snow, a pair of tied foxes. As he watched, the foxes either heard or sensed the approaching enemy, separated at once and made themselves scarce. I would imagine that the sudden fright engendered in a Highland fox by the least suspicion of the proximity of their mortal enemy would be enough to cause the immediate separation and flight of even the most lovelorn pair.

Many people will no doubt experience a measure of revulsion in reading of how the emphasis seems on destroying the hill fox, even at their mating time. I would personally go on record as saying that while I regard the fox as possibly our most handsome and absolutely vital wild animal, their damage potential, particularly to the lambs of the hill sheep flocks, but also in a certain measure to the young of roe and red deer, must always lurk in the mind of shepherd and stalker. And the fox, with every hand against it, continues to thrive in the Highlands—a survival of the fittest indeed, for the unwary or unfit seldom long survive the incessant persecution. Fox, eagle and deer: their lives and ways fascinate most of us, and stories are woven around them whenever stalkers forgather

After their period of bliss the mated pair, I believe, roam around together inspecting possible den sites, until when, at last choosing one, they may or may not do some cleaning-out operations so as to have it in readiness for the vixen's cubbing later on in the year. It is apparent that more than one sand-hole is cleaned out at this time yet not used. The digging instinct is strongly instilled in foxes, and it may be that they cannot resist digging a bit at any sand-hole they pass, before rejecting it as unsuitable. Certainly in any one year we may find many sand-holes (fox-cleaned-out) for every one used

Towards the end of January one year I lay watching a hind, her calf and her offspring of the previous year, a young knobber with as yet only knobs covering the pedicles from which his antlers would later sprout, come slanting up the hill towards me, the hill streaked with the snow wreaths left by a recent thaw. When about fifty yards from me the hind suddenly shied off to one side and sniffed at the ground, then walked around in a wary, stiff-legged half-circle, outstretched nose near to the ground, and sniffed again. Whatever it was that had alarmed her was below my line of vision as I lay flat and watched, and, intrigued as I was, I could only conjecture. The hind then continued to come up towards me, crossing a snow wreath so hard-packed that she slipped once or twice, her hard hooves barely marking it. Up she came, passing me only five yards away, stopping occasionally to peer fixedly but without comprehension at my motionless and only half-hidden form. At one moment she looked long and questioningly, but my absolute immobility reassured her and she carried on beyond me for a step or two and began grazing. Her calf following twenty yards or so behind repeated the performance, as also did the knobber, redder in coat and more ragged-looking than the dark, sleek calf. Their suspicions were shorter lived than those of the hind, and they soon decided that the object, though strange, was harmless. All three fed in front of me for some time, the hind on a patch of heather, rapidly pulling at the tips, the others on grassy patches. So near was I that I could hear them pluck the herbage. Eventually they fed over the hill top above me and I then found that the cause of the hind's sudden shying aside was a dead vixen lying on a snow wreath. The ground was rock hard with frost that day and so I left the dead fox on top of a green knoll, intending to bury it when the ground was softer. I returned a few days later to find the fox vanished without trace, undoubtedly, I believe, 'lifted' by an eagle. Between fox and eagle there seems much animosity, and I have had experience of more than one fox carcase eaten by eagles in the winter time, and have also seen fox feature as prey at eyries more than once. I know too, from a veteran keeper from an era when traps were often set for eagles, of how he went to visit a trap set in a known eyrie and found a fox caught in it. No doubt the fox had had his own reasons for visiting this eyrie. Certainly many eyries can be walked into, and I have little doubt that a bold, hungry fox might chance it if it thought it had the prospect of a helpless young eaglet in the parents' absence. There is also of course the scavenging aspect, the tempting smell of the prey which may be on the eyrie.

That mutual antipathy exists there is no doubt, whatever the reason.

The wildcat has an unenviable reputation for utter ferocity, a reputation zealously fostered by the more sensational writers on wild life, to whom indeed sensationalism would seem to matter more than fact. The few brushes I have had with wildcats have not led me to worry overmuch about the possibility of being clawed to death by them; in fact, as with every wild animal on the British list, the wildcat will invariably take refuge in flight, *if it can*, rather than take aggressive action when confronted by a human. An animal absolutely cornered, again of practically any species (except perhaps the badger who often prefers to 'lie doggo'), is a different matter. As a last resort here a rush may be made, apparently at its persecutors, a rush, however, which invariably will be made more with a desire to escape than anything else, but with potentially hurtful consequences should a collision occur.

I had a good illustration of this overwhelming desire to escape rather than be aggressive on the part of a wildcat when my terriers treed a large wildcat which had been raiding hen houses in my area one winter. The cat climbed right up among the thin, whippy branches at the top of a large birch tree, while my terriers yapped below, jumping and scrambling part way up the rough bole of the tree and sliding, blunt claws scrabbling frantically, all the way down again. I had no gun or rifle with me and so there was nothing for it but to climb the tree. Walking-stick clutched firmly in one hand up I climbed, feeling, I can assure you, far less at home when I arrived among the whippy branches at the top than the cat could have done. As I arrived within reach of the cat, crouched facing me, I paused to take a better grip of my stick, and as I did so, so did the glaring cat spring, but not at my face as it could so easily have done. No, instead it sprang for the topmost branches of an adjacent birch tree and, missing them, fell to its death below.

In my area of Inverness-shire, where mountain hares have long been scarce, the wildcat relied, in pre-myxomatosis days, mainly on rabbit. At that time rabbit trapping and snaring was a winter occupation for professional trappers, and their bag here invariably contained around a dozen wildcats each winter, trapped or snared as they prowled about rabbit burrows or along rabbit runs. This habit of following rabbit runs led to the very unusual death of a wildcat here. Returning from the hill one winter evening I saw something dangling in the upper branches of a birch tree above my path. This tree was below a wide green flat which at that time held a sizable warren and which in consequence was snared each winter. Going up to investigate I found a large dead male wildcat dangling from a branch high up in the tree, a wire snare tight around his stomach. Caught around the waist by the snare, he had pulled out the snare peg in his struggle to get free, but in doing so had drawn the wire agonizingly tight around his middle. Running away from the pain and from the bouncing, terrifying snare peg inexorably bounding after him, he had at last rocketed up the tree, ever the ultimate feline refuge. There he had apparently jumped from branch to branch until he had succeeded in wrapping the wire around one, and in jumping for another one had been jerked up short and, so hanging, died.

I was lucky enough to watch a wildcat stalk a rabbit one winter afternoon, and was intrigued to see that it was stalking entirely by scent. Absolutely concentrated on the stalk, it did not see my arrival as I sank slowly into a crouch to watch developments. Just as it seemed that the elongated, flattened form of the cat would snap into its final electrifying spring, the rabbit flicked around and into the burrow at which it had been sitting. The frustrated cat arrived at the burrow almost at the rabbit's tail and thrust head and shoulders into it. I almost expected it to follow the rabbit down, but it probably knew from past experience just how fruitless this was likely to be. Emerging, it looked around and, spotting me now, made off. Possibly the rabbit had noticed some slight involuntary movement of mine which the cat, engrossed in the hunt, had not, and this had caused it to vanish down its burrow; or perhaps the tension of watcher and hunter had communicated itself to the intended victim. Whatever cause, the cat was cheated of a meal, and I of seeing the successful conclusion of the wildcat's stalk.

Wildcats, unlike foxes, seldom eat carrion, even in the stress of winter's food scarcity, Their sharp but relatively small teeth are not adapted to tearing through deer hair or sheep's wool to get at the flesh below. I have, however, scared a wildcat from the carcase of a blackface ewe, and on investigation found that it had been skinning the head and eating the peelings of skin so obtained. Surely a very hungry cat to have been reduced to such meagre pickings. Since then, however, I have found this very characteristic skinning of the head, no other part being touched, on sheep carrion. Had I not actually seen a wildcat at this I should still be puzzled as to what did it.

While the wildcat population throughout the Highlands is probably not high, it may well be higher than many people believe. Expert in utilizing cover and, like the fox, mainly nocturnal, they have ample cover in the rocks and trees of their favoured river glens, while on the bare, higher hills they have recourse to the drier hill drains or burn banks. Skulking along these often heather-overgrown pathways they will see, hear or scent humans while themselves remaining unseen. I was once able to show some friends a three-quarter grown wildcat curled up, snugly asleep, below the heather fringing the bank top of a small burn down which we were walking. A few yards on either side and we would have passed the sleeping cat with no awareness of its proximity. How often in fact must we pass wild life by, unaware of its nearness. Stoats are not particularly numerous around my home, more particularly since the myxomatosis-induced decline of the rabbit. They seem, however, to have the faculty of popping up at odd times and in odd places, such as the one my Labrador bitch chased far out on the hill, and which escaped by diving into the deep waters of a nearby burn and swimming across with all the aplomb of one born to this element. Or the one which I found dead, as prey, on the eyrie of a golden eagle. Or the one in full winter white which a friend of mine bolted from a hole in the mesh of his outdoor meat safe. That enterprising individual had had a good meal of venison as a reward for initiative.

The stoat of course changes its coat in winter, a legacy, one feels, of the days of truly arctic winters, for nowadays, when more of our winter days are 'black' instead of white, it is often glaringly obvious in its yellow-white coat.

A track one seldom sees, in those telltale conditions of winter snow, is that of the badger. Many people still believe that the badger hibernates in winter. While this is certainly not true, badgers just as certainly do not like to venture from their warm setts in snowy conditions. In a long period of snow, however, hunger may be the deciding factor and then the badger too may show an unsuspected taste, perhaps of necessity, for carrion, its unmistakable broad tracks leading to a sheep or deer carcase, and thence, after dining, back to its sett. A nauseating diet to us; but Highland winters are long, and fresh prey is often scarce even to the omnivorous badger, which in summer can live very well on a plenitude of very small pickings, including beetles and earthworms.

Ivy, being almost the only greenery available in deep snow, seems very attractive to the deer family in winter, particularly to roe. Year after year one may see in snow the dainty cloven imprint of roe deer hooves leading along the same places in the river banks, skirting around wind-blown trees, down, at times, into cliff faces that would appear to daunt even chamois, there to browse on the ivy clinging to the rock.

Japanese sika deer, which were introduced to Loch Ness side about 1900, and which have adapted themselves remarkably well to our conditions, are also very fond of ivy in winter, and indeed are very like roe in their choice of habitat and feeding habits, seldom being found very far from trees. In size they are between red deer and roe deer, with antler formation and casting and renewal times as with the red deer. Their mating, or rutting time, is also more or less as that of the red deer. Stockier than the latter, they have a very prominent white caudal disc, reminiscent of roe, in their winter pelage, which is otherwise almost black looking. I have seen sika stags reared up, almost perpendicularly, on gnarled tree trunks, while they fed on the mosses or lichens on them. Neither red nor roe have I seen do this, though roe will certainly eat the mosses at the base of birch trees in winter, and I have seen red deer stags do likewise. Red deer will also take advantage of their superior height in reaching up and browsing on the thinner branch tips of birch and rowan, even rearing up on their hind legs to do so. Deer of course cannot *cut* with their front or incisor teeth, having no upper incisors but only a hardened gum pad against which the lower act. The branches being browsed, if at all thick, must therefore be cut by the cheek teeth, of which they have both upper and lower sets.

I have seen sappy branches of about three-sixteenths of an inch thick cut thus by red deer while browsing. This lack of upper incisor teeth leads to that characteristic 'plucking' motion as deer feed, the lower teeth cutting upward against the gum pad and the head of the deer throwing slightly forward and upward at each 'plucked' bite. This forcing upward of the lower cutting teeth means that the deer must have some resistance against it in the food species, particularly in the case of any hard substance through which the teeth cannot readily bite. This is what causes the distressing amount of damage in a turnip field which deer have raided, damage out of all proportion to the amount actually eaten. If a turnip holds fast in the ground it will be almost completely eaten, but if it comes loose at the first scrape or two of the deer's teeth, as so many do, it will yield to the deer's attempted feeding thereafter and so another *rooted, firm* turnip will be tackled and so on down the line, leading to a rise in the blood pressure of most farmers on the day after.

Winter into Spring

4

February

BEGINNING OF THE 'HUNGRY TIME'

Spring, with its burgeoning of new growth and new life, is always backward on the higher ground of the Highlands. Indeed throughout March and into early April those weeks, the aftermath of the Highland winter, with virtually all the previous year's growth seared and withered by frost and biting winds and still lacking new growth to replace it, test to the utmost the astonishing ability of Highland red deer to survive on minimal food and shelter.

Annually, under these conditions, Nature's age-old law of the survival of the fittest holds sway. The age groups under hill conditions from which most of the unfit come are the very young (calves of around nine months of age) and the ageing deer (fourteen–fifteen years upwards), with casualties in the first-mentioned group predominating. This annual mortality in the calves is a weeding-out process of any late-born, puny or weakly calves to the ultimate benefit of the species—natural selection exemplified. While it is always sad to see young life prematurely cut short it must be realized that Nature is never sentimental and seldom tolerates weaklings. I believe, however, that if hind numbers were pruned, in areas where they are *really* excessive, and selective shooting in getting rid of as many as possible of the poorer quality deer was practised, this annual calf mortality might be lowered significantly.

At the other end of the scale, the ageing deer, those whose cutting incisor teeth are worn down or non-existent and whose grinding cheek teeth or molars are ground to a polished concavity instead of their former saw-edged sharpness, will also fail to survive the rigours of scanty pickings. Without recourse to artificial teeth, good teeth are a prerequisite to survival in our wild animals; when the teeth begin to fail, prematurely or in old age, so does the end draw nigh. Seldom, under present-day Highland conditions, do red deer stags survive to make old bones, but hinds, in their antlerless anonymity, and with less predilection to raiding arable ground in winter, are another matter. I usually find a few aged hinds dead each spring, even though I make a practice of endeavouring to cull ageing hinds each hind-stalking season.

To me perhaps the most harrowing death in this annual calf mortality was that of a 'white' red deer calf which I first saw on 27 September 1960 while we were stalking the stags. Spying at a herd of hinds I could not, at first, believe my eyes when I saw the white calf lying in its midst. Its white head, seen through my glass, looked almost like the head of a Cheviot ewe. As the herd moved off later I saw that the mother of the white calf was a normal red hind.

On that September day the white calf was about five and a half miles from my home and from the Inverness to Fort William road, but in the following December I was surprised and more than a little apprehensive to see it right down on the low ground within sight of that road. Red mother and white offspring were now on their own, away from the herd and too vulnerable altogether so near to the road, with memories of roadside spot-lighting and deer-poaching still very much alive in my memory.

I made a practice for a few successive days of shoo-ing the motley pair away from the hill face above the road and out as far to the hill as I could manage; and at last it seemed I had succeeded, for when I next saw them they were again with the herd, not as high as in September, but on that herd's wintering ground some three and a half miles from my house.

Towards the end of February I again saw the white calf, still out on the hill and apparently fit and well.

Imagine my very great disappointment then to find on 24 March the white calf dead, immediately after a spell of fierce, squally and wet March weather, a spell of concentrated March venom such as always causes casualties among the weaker deer calves. It proved to be a hind calf of about nine and a half months old, unfortunately a rather puny one, and the difficult days of late March had proved too much for it. Two years before this I had seen a partly white calf here, but it had very obviously been a puny one and it also failed to survive the crucial period of its first winter.

These white or part white deer which occur here every so often are not albinos, but may well be descended from a strain of white red deer which were introduced to Glengarry about 1875. Certainly white red deer have recurred occasionally in this area, never more to my knowledge than one at any one period, but perhaps all the more conspicuous for that among the orthodox reds. In the early 1940's we had a white stag which travelled unmolested between the estate to our west and ourselves. He came to grief one winter, being found dead near the main road. It was believed that he had been hit by a vehicle while attempting to cross the road at night, not altogether an uncommon occurrence in the Highlands.

The next white beast I first saw in 1946, on a July day when I was home on leave from the Army, a striking sight, in the midst of a large herd of red deer. A hind this time and *at least* two years old, she was really white, no yellowish tinge apparent. I saw her many times in the years thereafter, sometimes at very close range, and she was a strikingly handsome creature with normally coloured eyes, lacking the pink of the albino. She lived on the east side of the ground and spent most of her summer on the high ground of a neighbouring estate, but at the autumnal rutting period, and in winter, she was usually on our ground. Many were the people I 'showed' her to and many's the time I spied at her myself while stalking stag or hind, and always she evoked interest and speculation with her handsome, unorthodox colouring.

In the winter of 1956 she was yeld, i.e. she had no calf at foot, and was consorting with a small herd of yeld hinds, eleven in number, well out on the high ground. She must have been at least twelve then, probably older, and was still a fit-looking, handsome animal.

The following year she had a calf at foot, but it was a terribly poor-looking calf and she herself looked in poor shape. I saw her that winter (1957), but then she disappeared and I have no doubt she died, from natural causes, some time in the spring of 1958. She could well have been about fifteen years old, a fair age for Highland deer.

In 1959 a cream-coloured young hind again occurred in the same glen, not as truly white as her predecessor, but white enough to be remarkable. She, however, disappeared shortly afterwards, just when it seemed we were to have a replacement for our old white lady.

And so to my white calf of 1960–1 and her premature demise. Latest occurrence in 1965 was the appearance of a parti-coloured calf, again in the same glen, with a silver-white head and silvered flanks. Should it survive to maturity we may again have a notable beast to look for and remark upon, with its assurance that the white strain is still recurrent.

It is in this spring period that I have seen most accidental deaths among deer, at times caused by their below-par condition, at others by their venturing into dangerous places in search of a 'green bite', or by a combination of the two.

By far the most common causes of accidental death in my experience are those involving wire. This, in the shape of rusty loops and old strands where fences once stood, and in the slack wires and, perhaps, dilapidated netting of ageing sheep fences, is potentially dangerous to deer and it claims victims every so often. A most strikingly gruesome tribute to the hardihood of red deer I saw in March a few years ago. A stag had been found dead by a colleague, and I was invited to go and view and photograph it. This stag had a length of multi-strand fence wire wound around head and antlers, a ragged-ended streamer of it flying from one antler top. Tatters of velvet which should all have been shed the previous September still clung to his antler tops. I am sure he became fixed in an old fence while rubbing some of this velvet off in that month, and after a Herculean struggle had broken the wire and pulled free. His struggling, however, had pulled the wire terribly tight around head and antlers and there was no getting rid of it. At the time of his death, some six months later, it had grooved deeply into one antler, while, even worse, it had worn a groove in the bony socket of one eye and was so tight around the lower jaw on one side that it had worn, with the constant jaw movement necessary in a 'cudding' animal, an inch deep into the living bone, and created a circular hole at the end of this groove. The agony this stag must have endured, in feeding simply to keep alive, baffles description.

More commonly, however, accidents involving wire occur in jumping a sheep fence, which in many areas deer may do as a matter of course in their daily movements to and from feeding grounds. To deer in good heart these fences present no difficulty, but when they are low in condition, as in early spring, they may often hit the top wire in jumping. Occasionally a hind leg may slip between top and second wire, thus tripping the jumper, whose impetus in falling forward can cause the wires to twist over and snare the leg tightly. To a human, endowed with reasoning powers, an easing *backwards* of the trapped limb would probably free it. To deer the ensnaring wire is an enemy which has caught it and which must be pulled against in frantic endeavour to escape. Sometimes they succeed, but too often one finds a hind, a calf or even a young stag trapped thus, dead. Occasionally I have rescued deer from this predicament, often at the expense of wire-nipped fingers, for I do not carry pliers to the hill. Most unusual of these fence accidents was that of a hind and a calf trapped within a few yards of each other, surely a million to one chance. Were they mother and calf, trapped together as they had jumped? Or were they quite unrelated victims trapped at different times? This must remain a matter for conjecture.

An ivy bush, enticingly green when all else was withered, claimed a Japanese sika stag as victim one year. In craning too far forward to snatch another tempting mouthful he had overbalanced and gone over the edge of the small cliff from which the ivy bush was growing, and had, most unluckily, caught the hoof of one hind leg below a thick root of the ivy while his haunch had jammed behind a young birch sapling. Had he fallen clear he could hardly have escaped death, but it would probably have been a quick one. As it was he hung suspended, grim fruit of the ivy bush, and there I found him days after his death.

In another year I found a red deer hind dead in somewhat similar circumstances. She had apparently poked her head in between two close-growing sapling birch trees, rooted on the very edge of a river cliff, either to rub her head up and down between them or to reach for a tempting bite beyond them. Whatever the reason, in her nearness to the cliff edge her feet had slipped presumably, and she had gone backwards over the edge, her falling weight pulling her head down into the vice-like grip of the 'elastic' saplings. She was very dead when I found, photographed and marvelled at her.

Generalizations on wild life, especially if dogmatically asserted, are always dangerous and fallible. A popular one, that red deer will *never* bog themselves, is quite definitely not true, though deaths from this cause are extremely infrequent, much more so than with hill sheep, who regularly fall foul of bogs. I have known deer calves to die in bogs, youthful inexperience leading to their downfall, but I have also had instances of stags bogging themselves. An ill-judged choice of a boggy pool in which to wallow led to the death of one stag. I found him when it was too late to be of any assistance. He had chosen a small, narrow and presumably deep semi-liquid hole and, entering it, had wedged himself in it and could not apparently find firm bottom to enable him to heave himself out. For this mistake he paid the ultimate penalty.

Strong, high and tight must be the fence, in this lean time of the year, which is to keep red deer stags from grazing or shelter which they consider desirable. I watched, incredulous, a few years ago, as two wide-antlered, big-bodied stags went *through* a netted Forestry Commission deer fence which had been erected only four months or so previously. It appeared absolutely impossible from my viewpoint about half a mile distant, this piercing of a seemingly impenetrable netted fence. When I went down to investigate I found that a gap had been opened, either fortuitously or by design, by apparently pushing up the uppermost wide mesh section of the netting away from the lowermost fine mesh rabbit netting and through this gap of *sixteen inches* high and *forty-four inches* long the stags had gone. It had been illuminating to see through how small a space a big stag *could* go, first wangling his head through, turning and twisting his antlers to clear them, when his body followed in one effortless motion. Undoubtedly it took a certain amount of initiative and know-how, for a third stag that day, younger and smaller in head and body than the first two, had given up after some abortive attempts to follow them. This passage I saw had been in use for some time before I had found it, as the trodden path and tufts of deer hair by it bore witness. Comically enough, there were in that fence three wide-open gateways without gates at that time; but red deer hate to use, or obtusely ignore, open gateways in a new fence. I have seen them run, maddeningly, right past such gateways when attempts were being made to drive them out of a recently enclosed area.

Characteristic of our red deer is their utter fearlessness towards water, river, loch or canal. Occasionally, very occasionally, accidents may happen when crossing rivers in spate; possibly a slip in raging waters too tempestuous to permit of recovery, and then successive stunning blows against rocks as the victim is carried away, end any chance of survival. Waters too deep to ford will be swum without hesitation. A friend of mine, at one time living in a house overlooking

Loch Garry, used to watch hinds regularly swim across to graze on the islands on the loch. If any hinds remained on an island for any length of time during the rutting season a stag would be sure to swim out to them. My friend's description was, 'They swim very light [i.e. buoyantly] in the water,' and so indeed they do. My notes for 1962 contain the following paragraph: 'A memorable sight today, sixteen stags in line astern, swimming the Caledonian Canal, west from Fort Augustus. Very buoyant looking, line of spine just visible above the water, head and neck held high and erect, no straining forward, no splashing of legs, just an effortless almost noiseless progression in which the first stags were emerging dripping, as the last were entering the canal.' This really was a most impressive sight and one which I would dearly have loved to photograph; but I was too far away, I had to be content with watching from a distance, and content I was to have watched such a spectacle.

Knowing of this aquatic ability of red deer I was amused when on holiday in Belgium some years ago to see a huge sixteen-pointer stag, with three hinds, confined in a small deer park. Confined, I say, but in fact while there was a high brick wall around three sides of their enclosure, on the fourth side was only a narrow moat. No Highland stag this, was my amused thought, but probably even a Highland stag would have little incentive to swim free if he was well fed and had his own private harem. Wild animals, and in this deer are no exception, seldom exert themselves for exertion's sake alone; invariably there is an underlying reason, usually understandable. Invariably too the line of least resistance is followed in their daily movements, which is why a regularly used deer path through the hills is usually a good line to follow, even to the best fording places on the hill burns.

Roe deer share this aquatic ability; they too have been seen to swim the canal and out to islands when they so desire. Those erecting deer fences should therefore remember that water is no obstacle to deer: neither river, canal or loch will stop them if the incentive is strong enough.

5

March

WATCHING WILD LIFE

Roe deer are always delightful creatures to watch, but these small deer, being adept at utilizing the cover of bracken and screening foliage, are not always easily watched or even seen throughout the months of summer and autumn when this cover is at its height in their woodland habitat.

In spring, however, while shielding leaf and undergrowth is yet dormant, much enjoyment can be had in watching roe, with loosely knit family groups, bucks, does and yearlings, still consorting with each other. Delightful as they look, human concepts of sentiment and chivalry do not apply to them, as was apparent in the behaviour of a buck towards the doe he was with on a snowy day in early spring. Just as red deer will do, so was this doe industriously clearing a small, roughly circular area to graze upon, with that lovely fluid sweeping movement of one foreleg. No sooner had she a patch cleared of snow, however, than this materialistic and unchivalrous buck would come across to her and make to butt at her with his velvet-encased antlers, so that she retreated and left him in possession of her clearing. This selfish manœuvre he repeated time after time as I watched.

On bright, frosty days in spring, with the wooded gullies and sheltered areas still chilled with overnight frost, roe deer like to get up into the sunlight, welcoming the returning strength in the sun's rays. Upon such a day in mid March, lying atop a heather-clad ridge just above the tree line, I lay and watched a group of roe: three bucks of differing ages and with velvet-encased antlers, consequently in differing stages of growth, and close by a doe with her twin yearlings, male and female, the male with barely discernible swellings rather than knobs on his head. Those of the group lying down in the long heather were almost hidden except for their heads, their jaws moving fast in the jerky roe-rhythm while cud-chewing, the motion ceasing at regular intervals while the chewed lump travelled lumpily down, then a fresh one up, the food-pipe at the front of the neck. A pair of buzzards, three hoodies and a pair of ravens were all, at one time or another, displaying and calling, tumbling and swooping above the peacefully cudding deer. At one moment the hoodies seemed maliciously intent on startling the deer, stunting and calling raucously, swooping and wing-flapping just above their heads. The roe deer missed none of this aerial *joie de vivre*, heads turned at intervals to watch, but were not in the least alarmed even at the hoodies' 'gallery' act. Birds and deer alike were enjoying the spring sunlight with its inherent promise of better days to come.

Deer with antlers encased in velvet are reputedly very tender of these; but that day a buck with half-grown antlers was decidedly vigorous in his use of them in dealing with an apparently urgent itch on the inside of one haunch.

The males of both red deer and roe of course bear antlers and cast and renew these annually, yet at widely differing periods in the two species. While the antlers of the roe encased in velvet are reaching full new growth in late March, those of the red deer are beginning to loosen and drop off. Roebucks in fact begin to cast their antlers, the mature, well-doing bucks first, in late November or so, new growth then beginning almost at once, so that some bucks may be in new hard antlers by the end of the following March, with a 'tailing off' throughout April and, indeed, well into May in the case of the younger, less forward bucks.

Red deer, as stated, begin to lose their antlers in late March, mature animals first as with roe, and with a long tail off in this casting, so that in May one may see a herd of stags in all stages, from older stags with new velvet-encased growth well advanced, to young stags still with old antlers intact. Late casting means late growth, and consequently the mature stags will be well ahead of their juniors, having their new antlers fully grown though still velvet encased by the end of July, and stripping or cleaning the velvet off, as the antler hardens, throughout August.

Cast antlers are undoubtedly difficult to find, and when one reflects on the area of heather-grown hill, intersected by drains, burns and peat bogs, over which the casting stags range at this hungry season, the fact is surely not strange. Seldom are both antlers dropped together, though I did once find a pair lying right and left, the width of a stag's head apart, just as if they had gently and simultaneously left their owner's head as he grazed. As the antlers grow loose any sudden jar or vigorous movement may shake them off. I have found antlers lying where stags habitually jump a fence, or at the base of a rubbing post, one curious one being a hydro-electric pole at the base of which I have found more than one antler. A shepherd once told me of seeing a stag, apparently much irritated by the looseness of his remaining antler, suddenly jerk his head upwards sending the loosening antler end over end high into the air, while a stalker in this area shot a habitual marauder on arable ground and saw his single loose antler fall off at the impact of the bullet on the stag's ribs.

One reason popularly put forward for the relative infrequency of antlers found is that the deer eat the cast antlers. This is a contributory factor certainly, for our red deer are extremely fond of chewing antlers, and indeed bone of any kind. I have quite a collection of chewed antlers, many chewed almost down to the basal coronet. This, in the case of, say, a thirty-inch length of antler with a varying number of points and a beam, or girth, of perhaps four and a half inches, is a striking tribute to the strength of the cheek teeth of red deer and to their keenness in chewing antlers.

The habit of chewing antlers and bone has been put down to a craving for mineral salts denied to Highland red deer in their largely acidulous hill grazing, and certainly when I inquired from a Belgian forester whether his red deer chewed antlers the answer was an emphatic 'No!' Better balanced feeding in more natural woodland habitat could explain this, or it may be that my Belgian informant was mistaken. While largely accepting the mineral-craving theory I must add that I have noted a distinct fondness for bone-chewing in the cross-bred Highland cattle here, and their largely arable grazing should surely not be significantly unbalanced.

Bone-chewing can lead to fatalities at times, a short, much-chewed, sharp-ended piece of bone wedging in jaw or throat leading to eventual death, while needle-sharp bits of chewed bone found at times inside shot deer must surely have led to internal trouble in due course. A hind was shot here some years ago with the skull of a sheep wedged tight around her lower jaw, just behind her incisor teeth. She had, while chewing at the dry skull, contrived to poke her narrow lower jaw through an eye socket and had been unable to free it afterwards. This can also happen with a pelvic bone of deer or sheep (a bone which has two oval shaped holes through it). Some years ago I saw a hind 'carrying' one of these pelvic bones on her lower jaw; but chances are that she got rid of this encumbrance, for I never saw her again, though I looked for her. A tame hind I had at one time would chew dead mice or small birds, while a colleague once watched a red deer hind walking nonchalantly along, chewing meanwhile on the dried, furry skeleton of a mountain hare. It seems in fact that red deer tastes in chewing are catholic, for I have watched a red deer hind, on the hill, chewing with every appearance of satisfaction a length of alloy tubing she had found; and I have had the best part of a pick handle, which I had left by a path on the hill, chewed away by deer, the chisel marks of their teeth unmistakable on the remnant of the shaft.

Japanese sika stags cast their antlers at much the same time as red deer, but whether they are addicted to antler chewing I have so far not been able to observe.

'Jap' deer will at times lie very close in cover, if they believe they are unseen, trusting to stillness rather than flight, and I had an interesting example of this in March one year. I was coming home one evening through a straggly birch wood when I almost stepped on top of a young Jap stag. He was lying in the sparse cover of an isolated clump of longish heather, stretched flat out, head extended along the ground, only the line of his back, tips of his ears and his shortish antlers really visible. The wind was blowing from directly behind me and, to be truthful, my first thought was that here was a dead beast, until I saw the ears twitch slightly. I began, half crouched now, to reach slowly for my camera, and as I did so his head shifted slightly, ears flicking. Camera ready, I looked through the viewfinder; no picture thus, I thought (how foolish I was), and so I 'clicked' with my tongue hoping for a 'head up' photograph. The stag's ears flicked, that was all. I 'clicked' again. No response! I whistled, I grunted, I barked in crude imitation of a red deer hind. I went through a fair repertoire of noises calculated to alert the dullest of deer. Absolutely no response. Apart from the seemingly involuntary ear movements the Jap remained motionless. I straightened up slowly, meaning to take a photograph anyway of the Jap lying doggo. As I did so, in one incredibly swift motion (I really cannot even recall seeing him *get up*, my first impression remains of him running off), the Jap was up and away. Finger on the release button of my camera, as it had been for five tension-packed minutes, I did not get my photograph. I had been completely outsmarted by the 'camera shy' Jap.

One would imagine a Jap sika stag, even a young one scaling some 80 lb. perhaps, to be over-large prey for a fox to aspire to, yet a year or two ago the wife of a colleague, living in an even more isolated house and higher in the hills than mine, *saw* a sika knobber being attacked with quite definite purpose by a large dog fox. She had heard, about seven-thirty one spring morning, what she assumed to be the barking of a dog near to the house. Her husband, out on a deer-poaching alarm the previous night, was still abed, and, hesitating to wake him unnecessarily, she went out herself to investigate. To her astonishment she saw the sika knobber penned, at bay, in the angle where two lengths of deer fence met, while the fox, running from side to side in front of it, kept it penned, now and again springing at its throat, or crouching in front, barking at it. The young sika was all too obviously panic-stricken, and though temporarily holding his own there is little doubt but that a determined thrust home by the big fox would have brought him success. My colleague's wife, with no thought in her head but to succour the young sika, clapped her hands loudly and shouted. At this rude interruption the fox hurriedly departed as also, a moment or so later, did the young sika. It was only then that my colleague's wife realized that if she had aroused her husband he could have essayed more positive action. What he said when she did wake him and tell him I will draw a discreet veil over. This young sika could well have been a weakling, low in condition in the crucial early spring period, and the fox was certainly a very large, powerful dog, yet it does show what a fox, possibly also in stress of hunger, will tackle on occasion.

At this time of the year, when deer are about at their lowest ebb, the real weaklings, those in fact probably marked for death anyway, are definitely vulnerable to the bolder predators. One March morning, as I walked out through a wooded river glen here, I came quietly around a bend and found myself within fifty yards of a golden eagle astride a dead deer calf, the hair of which littered the ground about it. The eagle rose and took wing at once—rather heavily, as it had obviously had a good meal. The probability here was that the calf had been dead before the eagle found it as the carcase was definitely 'high'. On another occasion, however, also in the glen banks, the probability was that an eagle had killed the puny roe yearling I found part eaten, with the plucked hair characteristic of the eagle's feeding habits scattered around it. Seldom, however, do any carcases of deer which have succumbed in the harshness of a Highland spring remain long before being found and eaten; fox, raven, eagle, crow and even badger will avail themselves of this carrion. While spring is a lean time for the grazing animals it is a time rich in pickings for those not averse to carrion, and only a clean-picked framework will eventually mark the demise of deer or sheep.

Both golden eagle and raven are early nesters, and both normally favour cliff sites, though I know of a large Scots fir, in splendid isolation among the trees of a birch wood, which in one year held the huge nest of an eagle. A heavy snowfall in the winter immediately following the appearance of the nest built up such a weight of snow on the broad platform of the eyrie that the sustaining branch broke, and the eagle never returned to the tree.

The popular conception of an eagle's eyrie would seem to be of a nest built on a fearsome, beetling crag whereon one must risk life and limb to get near. In actual fact this is more often true in the case of the raven; quite definitely, I believe, the more intelligent bird of the two, which in my experience often builds a nest in fearsome situations. A pair of ravens resident here use the same site year after year, on a nasty steep cliff with a river foaming over huge rocks at its base, a treacherous cliff, always greasy and slippery, which, because of its situation, seldom sees the sun's rays. The actual nest is sheltered from above by a bulging overhang, from below by its outjutting ledge, and on each side by shielding buttresses of black, sunless rock. By craning out dangerously at one side, with a firm grip on a slender, cliff-growing sapling rowan tree, one can make out its outer edge, and when the young are well grown and moving on the nest glimpses can be seen of them if they come to this edge. Otherwise the only viewpoint is from directly across the river at a distance of more than a hundred yards, an ideal site, one can only think, for a bird trying to escape persecution.

I had an intriguing insight into raven affairs one spring. Attracted by the sound of a raven's continual calling above some birch trees, and seeing the bird constantly circling over one area, I went to investigate. I found a hen raven dead below the birches, at that time of the year the bare area of skin on her lower belly, the brood-patch, identifying her. I assumed that the hen must have eggs already laid in the usual nest, and decided that there would be no young ravens there this year. I could not have been more mistaken. Two days later I saw three ravens circling above the nest cliff, and on the following day no less than seven sable birds circled and called over the area. Had some corvine matrimonial bureau sent along half a dozen eligible females in answer to an S O S from the widower? Jesting apart, the fact was that somehow, at a time when ravens are paired off at their nesting areas, these spare birds appeared. Were they in fact summoned by the surviving raven, and if so how? He did, in any case, select a mate from the six, for two birds remained at the nest cliff and in due course young ravens *did* leave the nest that year.

In contrast to this inaccessibility of most raven sites, the site of an eagle's eyrie, popular ideas notwithstanding, is nearly always relatively accessible to anyone reasonably nimble of foot and with a head for heights. I know of the sites of four pair of eagles around my home, and as all eagles use alternative sites, these eagles have altogether a total of sixteen sites among them. Of the sixteen, only two are inaccessible except by rope. Both belong to one pair; but this inaccessibility would seem to be accidental rather than premeditated, as the third site of this pair is easily accessible. One feature peculiar to most eyries is the shelter of an overhang above the nest ledge, and curiously enough most eyries have a rowan tree in close proximity. Rowan trees of course (their alternative name of mountain ash being very apt) will grow out from the merest fissure in a cliff, thereby escaping the cropping teeth of deer and sheep, and so the proximity of rowan trees to eyries is perhaps not really remarkable. Many eyrie sites have been in use for generations, and the lower layers of such eyries are truly impacted into a solid mass rather than of individual sticks. Large sticks are often in the huge structure of an eyrie (I know of one with a height of structure of 6 ft 3 in. when last measured). A stalker I know well told me of how a shepherd once recovered a horn-handled crook, which he had lost on the hill one day, from an eagle's eyrie later the same year. Spying at the eyrie his eye had been caught by the horn handle of his crook sticking

A sika hind (*left*) and calf in winter. Compared with red deer, they have shorter and more thickset bodies; like roe deer they have the white caudal patch.

The white calf, found dead at the age of nine months.

A weak deer calf, unable to cope with the strength-sapping conditions of a late snowfall. Photographed in March.

A herd of startled red deer stags jumping a fence.

X

This stag, probably weak in spring after the winter hardships, trapped a foot in the wires of a stock fence; it provided food for the vixen and her family described in Chapter 7.

Four big stags, antlers in full velvet, crossing a hill burn at the end of July. Red deer have no fear whatever of water and will cross river, loch or even the Caledonian Canal when they so desire.

Roe doe in winter coat showing white throat patches. This winter coat retains its sleekness until May in most cases.

A roe doe licking the new sprouting antlers of her last year's fawn.

XI

A roe family lying together in early spring. Bucks usually leave the family party when they clear the velvet from their antlers, living a solitary life until mating time in summer.

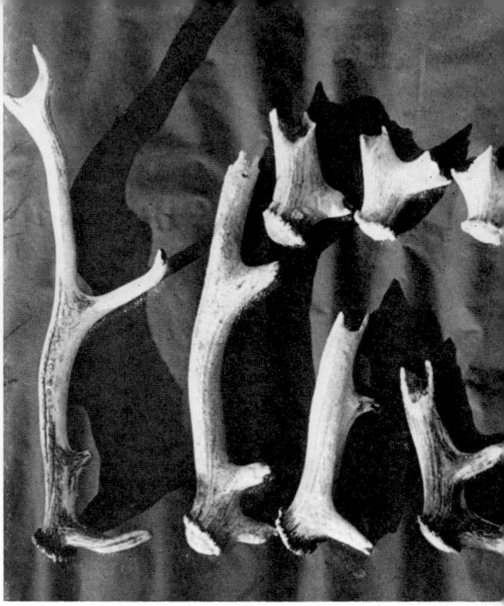

Red deer chew antlers cast by the stags, perhaps because of a mineral deficiency in their diet. A hind, and examples of chewed antlers of red deer.

Japanese sika deer, a hind, her winter coat already looking shabby in March.

The golden eagle's eyrie is often accessible.

XIII

The hen eagle brooding and—later—with two one-week-old eaglets in their eyrie.

Red deer stags which have recently *cast* their antlers; stags begin to lose antlers from the last week in March in the Highlands. Then, in April, the new antlers begin to sprout, as on this mature stag.

When the first foliage begins to appear red deer may rear up to browse on the tempting greenery, as this hind is doing.

A woodcock sitting on her nest in old bracken—a masterpiece of mid-April camouflage.

Hen ptarmigan sitting on eggs in high moorland grass by a peat hag (June).

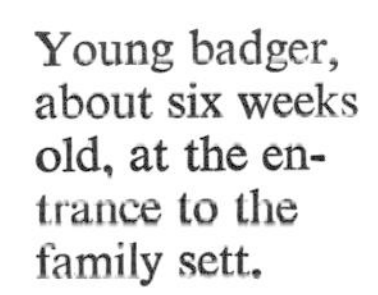

Young badger, about six weeks old, at the entrance to the family sett.

A badger on nocturnal prowl.

XV

The fox indulges a habit of chewing the tail off a lamb carcase.

The end of a vixen, bolted by terriers from a cairn, and shot.

from the nest structure, and so his cherished crook was recovered in a way never anticipated.

When an eyrie is to be used in any one year usually about a foot of fresh structure is added, large tufts of heather being mostly used. A very shallow cup in its centre, lined mainly with great wood rush, supplemented with moss and dried grass, ultimately holds the eggs, the normal clutch being two, though one is quite common in my experience. The nest being invariably a very broad structure, the egg (or eggs) incurs little risk of rolling out, as at first sight may seem possible in the shallowness of the actual cup. By the end of the long nesting period (forty-one to forty-three days' incubation followed by ten to twelve weeks of fledging time) the foot of fresh structure is beaten and trampled flat, rusty-red brown and withered, often with its outer edge apparently dangerously sloped away.

The eagle will usually lay about the third week in March, and I have watched one sitting on 14 March, who probably had not laid, however, but was forming the nest cup. This was one of the two inaccessible eyries referred to which I could only spy from a distance. The bird here was constantly resettling sticks on the rim of the nest as she sat, and after a few moments of this she left the nest and flew out of my sight.

Without real experience of eagle eyries one is apt to assume that, by reason of their cliff sites, an unobserved approach is impossible. This, however, is far from the fact. Practically all the eyries I know of have at least one blind spot, and in fact all my eyries are so broad of structure that even a huge bird like a hen eagle, when sitting in its centre, cannot see over its edge so that an unobserved approach from directly below is often feasible. At one eyrie I have often approached from below, worked quietly up to one side and been able to lie above and watch the hen as she sat, unsuspectingly, below. When near the end of the incubation period, the eagle will sit very tight indeed and is then *extremely* vulnerable, unfortunately, to those unscrupulous enough to plot her downfall. I had a friend with me while visiting an occupied eyrie one year, and we approached from below without any attempt at concealment. It was a steep pull up to the base of the eyrie cliff, and we were glad to sit below it and, quite sure that the hen had seen us and left long ago, exchange a few words about the climb and the possibility of seeing an adult eagle. As we speculated thus the hen, who had been on all this time, rose off the nest and flew up the glen, leaving us speechless for the moment.

At another eyrie some years later I actually spent some time, on a dull rainy day, shouting below it in an attempt to see if it was occupied without actually climbing to it. Nothing stirred, however, and I decided I had better check on the 'empty' eyrie. It was not until I was on a level with, and in sight of, the eyrie that its occupant took off, for it *was* occupied. Working my way across to it, the reason for the hen eagle's reluctance to leave was obvious: one of its large rough-textured eggs was on the point of hatching; through a ragged, starred, sixpence-sized hole in its thick, discoloured shell I could see moving the black beak of its lustily cheeping occupant. I left quickly and returned a week later to find both eggs now hatched and the carcase of a plucked grouse beside the two downy eaglets.

This eyrie I visited at weekly intervals (though it entailed a three-hour journey each way) throughout the fledging period. The eaglet last hatched lagged very far behind the first, and ultimately died at about seven weeks old. It was quite definitely not actively ill-treated by its nest mate, but it was almost certainly denied its necessary share of the prey being brought in to the eyrie. I believe it had survived so long because, up to about six weeks of age, the eaglets are actually fed by the hen, who tears up the prey and is probably more or less impartial in feeding each eaglet. After this age, however, the eaglets are expected to tear up the prey themselves; it is then simply left on the eyrie for them to deal with. This period was crucial for the smaller eaglet, and here it had coincided with a week of scant pickings (a week of constant low mist and consequent bad visibility, conditions all against a high flying bird of prey dependent on good vision). It is fairly common, actually, for one eaglet of two hatched to die before fledging, yet I have also seen two reared more than once. I believe it depends on the availability of prey in any one year: in a year of plenty two eaglets will fledge; in a poor year if one eaglet is significantly weaker than its nest mate it will succumb.

The larger eaglet left this eyrie at exactly ten weeks after hatching, and weighed 9¾ lb. then, from 1 lb. at a week old. She (for I believe it was a female) was obviously a strong, forward bird, for it is sometimes nearly twelve weeks after hatching before an eaglet leaves the eyrie.

6

April

NEW LEAF AND FOX DENS

APRIL, in a Highland deer forest, is seldom the Lady Bountiful she may be in the kindlier south. Much of the higher ground is often still shrouded in snow in early April, while even the lower hills may still be speckled with hard-packed snowdrifts. When these drifts do eventually vanish their positions will remain marked for some time afterwards by the brighter green of the heather which had been below, evidence of the protection afforded by the snow against the 'scorching' of frosts and withering easterly winds. It will be at least mid April before we can expect to see that lovely, delicate green nimbus, a shimmering mistiness rather than a tangible reality, enshrouding the earliest birch trees, and it will be well into May before the last of the high ground birches loses the nakedness of winter. It may well be, too, early summer before the prevailing withered, sterile aspect of the hill grazing changes from the sere and yellow to that greenness which ensures full bellies, with minimum effort, to deer and sheep.

April, nevertheless, is the month of renascence, harbinger of better days, implicit with promise of hardships coming to an end, a month when life becomes renewed in many species.

Golden eagle and raven will be brooding eggs and feeding young respectively. Blackcock may be heard 'coo-rooing' in the stillness of early morning, a lovely, soporific welcoming of spring which has a ventriloquial quality, making it surprisingly hard to pinpoint. Woodcock will be sitting on eggs, living masterpieces of camouflage, among the red-brown of yet-withered bracken in the birchwoods of the glens, and by late April many pairs of grouse will have full clutches.

The woodcock and the grouse will need all their camouflage, for in these same wooded glens hooded crows will be sitting on eggs, the bulky nest still obvious until the full leaf comes. It is a very dilatory or unobservant stalker who will let slip this chance of dealing with this entirely odious bird before she is successful in propagating her rascally race. The hoodie in fact is the one bird in which I cannot find one redeeming feature. Weaklings of any kind on the hill are immediately victimized by the sharp-eyed patrolling hoodie, eyes and tongue of weak young lambs or struggling ewes being removed while the victim yet struggles, while fence-caught or dying deer may be similarly operated on, alive. A shepherd came home from the hill here on one occasion after having found a weak ewe lying on her side, and, being in a slight hollow, unable to regain her feet. Her uppermost eye socket gaped emptily up at him, bloody and raw. He had lifted her and tried to restore her circulation, but she was too weak to stand and so he left her lying while he finished his round, intending to take her home when he returned. When he did get back an hour or so later it was to find her other eye gone. Little wonder his loathing of hoodies was redoubled thenceforth.

Eggs of every description of course are grist to the hoodie's mill in spring time, from the small song birds through domestic poultry 'laying out' to even the larger birds of prey. I have seen only sucked eggshells in the hoodie-ravaged nest of a buzzard, and have known two young buzzards to vanish while still at the grey-white, down-clad stage, while near by the four young hoodies in a tardily discovered nest waxed fat and flourished exceedingly—until found. I have seen the sucked shell of a sparrowhawk's egg lying near the nest tree, and the single egg of a golden eagle once vanished in circumstances which led me to suspect the hoodie. A colleague once told me of finding, in the not so distant days when every bird with a hooked beak was *persona non grata*, a merlin's nest with a full clutch. He had returned next day with a colleague in the hope of stalking the discovered nest and shooting the merlin. They found instead an empty nest and formed the rather ingenuous theory that the merlin had shifted her clutch to safer quarters. Attractive as this theory is, I think it far more likely that a hoodie had seen the stalker discover the nest and immediately upon his departure had taken advantage of the merlin's continued absence to despoil it. The eggs of small song birds usually vanish without trace, though once the egg of a yellow-hammer was found in a hoodie's nest, while depredations on the eggs of game birds are too well known to need emphasizing here.

At one time, in those more spacious days when Highland estates were not as understaffed as they are nowadays, the traditional and highly successful method of dealing with discovered nests was to build a hide (in Highland parlance, a 'bothan') within shotgun range of the nest. This was then left quiet until it was judged that the hen was sitting tight, usually about the end of April. Two stalkers would then go to the bothan, one, with his gun, being shut in by his colleague, who then left in as ostentatious a manner as possible. In due course the adults would return, usually singly, to the nest, and often both would thus be accounted for by the concealed stalker. A rather remarkable yarn illustrating this method was told to me by a retired stalker. He had been left in a bothan and, shortly after his colleague had left, a hoodie had come in and sat on a branch near the nest, to fall to his shot. Some hours later, cramped and chilled, the stalker decided to quit, in the belief that he had shot the hen and that the warier cock was not going to show up. Leaving his conceal-ment he decided to put a charge of shot through the nest. He was absolutely dumbfounded when, at his shot, the hen hoodie tumbled, in a flurry of wings, from off the nest. He vowed that she could not have come in unseen *after* the cock, the first bird, had been shot, but must have been sitting tight, unsuspected, when he had been put into the bothan and had continued so even at the shot which had killed her mate.

Such behaviour is of course exceptional; seldom does the wary hoodie sit *so* tight, although one very stormy day I was once half way up a tree before the hoodie flew off her nest. Nowadays, most stalkers being single-handed, the bothan system is seldom used, but one can still often get a chance at the hen hoodie by quietly approach-ing a previously discovered nest when she is at the sitting tight stage, preferably on a stormy day.

Among the mammals of the deer forest, badger, fox and wildcat will probably have young by mid April. The badger setts, where they are in sand (some Highland badgers make use of rock cairns), will be obvious by the huge flattened ramp of old bedding and sand built up in years of annual spring cleaning, while trails of fresh bedding, dried grass and bracken will lead down the sandy tunnels. Badgers are increasing nowadays throughout the Highlands, but being entirely nocturnal in their habits one sees much more of the signs of these nocturnal activities than of the animals themselves. Scrapes and rootings in the vegetation, the turning over of dried cowpats in search of beetles and worms, and in autumn the raw, clawed-out devastation of wasp or wild bee nest: these afford mute evidence of the presence of the badger. I remember well one April evening when, passing by an old-established badger sett on our way home after a long day looking for fox dens, we decided to linger a while and watch. As the gloaming deepened a long-snouted, black-and-white-striped head broke the black circle of one entrance. This way and that the long snout probed, testing the evening air until, satisfied, its owner emerged, to merge in fashion almost uncanny into the greyness of the dusk. Across to where we sat, in the cover of a shallow gully, we sensed rather than saw it come, until a traitorous puff of wind gently stirred—cool fingers playing on the back of my neck—and sent the badger scurrying for the safety of its sett.

At times the sand-hole den of a fox may also have a conspicuous pile of fresh sand at its entrance, thrown out in enlarging it to the vixen's liking; but there will be no bedding mixed with this, and it will not have the ramp-like appearance characteristic of an old-established badger sett.

The wildcat, in my experience, invariably 'dens up' in rocky cairns which show little outward signs of occupancy, though I have on occasion found betraying evidence in the carcases of shrews and voles, buried just below the surface of the 'doorstep' of an occupied den.

With neither badger nor wildcat have I any quarrel (except in the case of a 'rogue'), but in the spring quest for fox dens we sometimes contact, unavoidably, these other earth dwellers.

It is in late April that the lambs of the hill sheep begin to be born, at a time very convenient to the needs of the hill fox in feeding a growing litter with growing appetites. This then is the time of the year when vulpine depredation is most serious and apparent, and although some of the lamb carcases seen strewn about the noisome entrances of a fox den may well have been picked up dead, many are undoubtedly killed by the fox. Much of the Highlands, the west and north-west in particular, affords very meagre fare in the way of natural prey nowadays, and a fox, which in feeding only itself can exist very well on a variety of small pickings in summer and on carrion in winter, must have larger prey in feeding a litter of five or six growing cubs. Litters vary a lot in number; I know of a vixen killed while raiding a hen-house who had fourteen unborn cubs inside her. At the other extreme I have known of only one cub. The average number, however, will be five, and to meet their needs the fox often turns to lamb killing, and, having discovered what easy meat this is, may become a confirmed lamb killer. It might well be that were those other vulnerable hill younglings, the red deer calves, born in April instead of in June, this very real menace to lambs might not be so great; as it is the lambs must fill the bill of fare. Measures therefore have to be taken to deal with this situation throughout the Highlands each April, by finding and dealing with fox dens before the almost inevitable lamb losses become too large. For this work we rely on terriers, going on our first round of all the known den sites about mid April. Cunning and resourceful as the fox undoubtedly is, it has the blind spot of using the same dens, though not always in successive years, year after year. Most estates have in fact traditional fox dens or cairns reaching back for generations of stalkers, often with appropriate legend. This is not to say that new dens are never brought into use by the fox, simply that year after year a round of the traditional sites will usually bear fruit. As with litter size so does the cubbing time of the vixen vary, quite widely at times, but the peak time is about the first week in April, and we believe that the vixen is usually to be found in, with her cubs, for the first ten days of their life, a stage when they will be small, vulnerable and sightless. The dog fox will bring her food at this time. If therefore we can discover her den and enter our terriers, the vixen will often bolt and is shot as she does so while the terriers dispatch the very young cubs at once. Distasteful as this battening on maternity is, it is to prevent vulpine battening on maternity, and is as humane a method of dealing with the Highland fox as any, far more so than gin trap or wire snare. To two or three skilled shots disposed strategically, with the acquired knowledge of years, around a fox den, a bolting vixen affords little problem and death comes quickly.

For the terriers we use I have a great respect, these dauntless, diminutive warriors, cheery, canine 'pot-holers' whose path of duty takes them into a dark labyrinthine underworld. Forsaking fresh air and the light of day they have to thread their way through narrow, rib-pinching, rocky passage-ways, the Stygian gloom hiding many pitfalls, or down constricting tunnels of sand, with eyes, nostrils and coats full of the clinging stuff. Or, worst of all perhaps, a squirming through the absolutely chilling muckiness of wet peat runnels from which they emerge black and dripping, so caked with wet peat as to be filthily unrecognizable. And at the end of their underground searchings—the snapping jaws of a long-toothed vixen, holed up you may be sure, so as to have the advantage in her defiance, temporary though this may often be. In many cases the vixen undoubtedly is in such a narrow place that the bulkier, less 'elastic' terriers cannot really get to grips with her. Slit ears and neat punctures on muzzle and top of head may be inflicted on the terriers, but I have never personally seen substantial damage inflicted by a vixen, nor is it often that a vixen is killed underground. Maternal instinct in a vixen would seem to me to have been popularly over-emphasized, for often a vixen, even though well placed for defence, will bolt after a varying time from the terriers' entry. Presumably her nerve cracks and she prefers to run for it rather than stubbornly defend her cubs. I have sometimes thought that in thus bolting she may believe she will draw the terriers away after her from her den, like the decoy display of some birds when their young are discovered; but perhaps to ascribe this motive to her perfectly understandable flight is trying to read too much into it. I do know, however, that a vixen does not, as I had at one time always assumed, always take up a defence stance *between* her cubs and the attacking terriers. This was made evident to me some years ago when I dug out a vixen with young cubs of about a month old, who had intended to bolt, but had been approaching the single entrance/exit just as I had unfortunately decided to peer down it to see if my terrier was emerging. Instead I was just in time to see a long, slender white and whiskery muzzle, black-tipped, appear inquiringly around a bend in the hole. Petrified, I watched, but nothing more appeared, and in a moment the inquisitive muzzle melted back into the gloom of the hole. I knew by hearing my terrier baying the vixen that this must be a very shallow den (it was in fact an enlarged rabbit hole), and so I eventually dug it out. I had wondered how, in a single-entranced hole, the vixen had managed to get out past my terrier and then back to her former stance without encountering him. When I had finished digging the den out, everything was made plain. The single tunnel led in for a short distance *and then forked*, and the divergent tunnels had gone widely right and left to meet again, completing a circle, roughly, at the rear of the den, a circle from which a further short single tunnel led off to the rear. At intervals around this relatively wide circle there led off some very narrow rabbit 'pipes', into which neither adult fox nor terrier could penetrate. In these pipes were scattered the cubs, while the vixen was holed up in the short, narrow tunnel leading off the rear of the circular tunnel, her head filling its narrow entrance. Her cubs were therefore nearer the entrance than she, not behind her in a last-ditch defence. The circular tunnel narrowed as it went around to the rear of the den, and my terrier had been unable to get right in to the rear offshoot, though near enough to keep him trying, and while thus engaged the vixen had slipped out the other arc of the circle intending to bolt, but had changed her mind and retreated again on seeing or sensing me at the entrance.

There *are* times of course when a vixen will *not* bolt. This may, in some cases, be because the terriers are in her way, and we always try, as a last resort, the expedient of leashing the terriers away from the den's entrance, after we get them all out, and waiting quietly for a while. Very often then, sometimes within minutes, sometimes more than half an hour later, the vixen will bolt. The first vixen I ever saw bolt, many years ago now, got away simply because she bolted immediately after the two terriers we had in came out. My veteran colleague and I were fully, and foolishly, occupied, guns laid aside, leashing a terrier apiece when, a blur of red, the vixen left the hole. I was very green then, but my experienced colleague should have known better. Never again have all the guns been immobilized simultaneously at any den I have been at. Cases like this of course go to augment the numerous legends of vulpine cunning, but I believe that the vixen's instantaneous bolting (while the enemy had its hands full) was purely fortuitous. Fortuitous or no, it saved her skin that day.

The scales are not always weighted against the fox, for yearly there are casualties among the terriers used at the den time. Of late years I have lost one Cairn terrier, swept away and drowned while attempting to cross a snow-swollen river behind me, and a cherished, staunch Cairn bitch who failed to emerge from a sand-hole which in the two previous years had held a vixen but on that occasion, we belatedly discovered, held badgers. An ever present risk is that of a terrier sticking in a 'cairn' den, a den, that is, in the huge jumble of rocks which cover many Highland hillsides. Cairns are much favoured by foxes in some higher ground areas where sand-holes are few, and are always potentially hazardous to terriers. They may jump down into a slit in the rocky bowels and be unable to jump back, or literally jam themselves into a narrow cleft in an attempt to reach a cub. Understandably, though unfortunately, the keener the terrier the more risky is its existence at this work. Sometimes Herculean labours, in digging down, often with bare hands assisted by a walking-stick, can free a terrier, *if* the rocks are loose enough and not too large. I worked all of one afternoon in a persistent downpour to free my terrier who had absolutely jammed himself, luckily not too deep, in trying to get at a cub. When eventually, filthy with peat, finger-nails all broken and finger ends raw, I got to within sight of my terrier and freed the last slab which held him, he made a dart forward *into* the cairn and grabbed the cub which heretofore the constricting rock had denied to him. A sight I shall never forget in this impromptu rock-mining work was that of the two feet of a colleague kicking and waving out from a vertical body-constricting shaft, which hid all else of him, as he strove to free yet one more rock to rescue his trapped terrier. Our method at this last stage of that particular day-long rescue was for him to descend head first into the vertical shaft we'd opened, and to then grasp a stone at its bottom. Another colleague and I then grasped a leg apiece and drew stalker and stone bodily out. Thus, stone by painful stone, we ultimately freed the terrier which had been in for almost twenty-four hours.

Needless to say, such work among huge, unstable rock is fraught with risk, yet with a cherished terrier's life at stake this risk is often undertaken. In many cases, alas, in very deep cairns in immovable rock, all efforts may be in vain; one can only mourn when all hope is gone.

Though I am not inclined to think highly of the maternal instinct of the vixen, occasionally, very occasionally, strong and undeniably heart-stirring evidence to the contrary turns up, as when a bolting vixen is shot, and proves to have been carrying a very young cub in her mouth. I have in fact only ever heard of this three times, from *reliable eye-witnesses*. In two cases colleagues with a lifetime's experience had each seen it occur once. In the third case a relatively inexperienced chap had also seen it. He may never see it occur again.

As the cubs grow older, however, both vixen and dog fox seem to show more strongly the parental ties. In the case of a belatedly discovered den the vixen, with cubs older and stronger now, may well be out during the daylight hours. The procedure often followed here is to deal with the cubs first and then to wait out, all night if necessary, posted strategically around the den. The best chance obviously is when one has the element of complete surprise and an unsuspecting fox is bagged as it comes openly in. Occasionally things work out like this, but more often the adults, one of which may have been lying quite near the den all day, are suspicious and may not offer a chance before darkness temporarily ends the ambuscade. Both parents, however, may exhibit extreme perseverance in their efforts to approach a den *even* when they *know* there are humans about.

When the darkness does halt operations a return to the den is made, and a fire is usually made to brew tea over and to instil some warmth into chilled bodies. An hour or two of attempted sleep follows in as much comfort (*sic*) as the crannies around a den may afford, while every so often in the night one may hear the scalp-prickling scream of the vixen as she calls to unresponding cubs. Before it is daylight a move is made, leaving one 'gun' at the den, out to positions from where it is hoped to get a chance at an overbold fox when the light strengthens. Foxes are as individualistic as humans, and so at times we succeed in bagging one or both adults, and at times we don't. These night operations, however, do succeed often enough to keep us trying, especially in the case of a pair known to have done much damage which they may well repeat the following year if not brought to book.

Nights spent out in the brooding immensity and solitude of the darkened hills, where at times one can feel absolutely alone in the world, vary enormously, but are always fraught with tension and tingling expectancy when in the gloaming or in the weak light of dawn one hears the eldritch scream of the vixen and strives to pick her up, eyes straining, in the wilderness of rock and heather. Nights of frost when it seems one cannot endure any longer the cold of the interminable night, but one does; or of heavy rain, as the one I spent crouched all night in the sandy slit trench of a dug-out sand-hole, my coat spread above, and waited longingly for the first signs of daylight. That night proved fruitless; we heard the vixen, but saw her not at all. Small wonder, in that weather. Another fruitless night, in terms of adult foxes bagged, was, however, enjoyable in the tangible tranquillity of a balmy late April night far out in the hills, the only sounds, night long, those of the wild life around us. That night was overcast but not pitch black; indeed the moon was almost full, but hidden all night above a grey cloud layer. In the semi-darkness the gaunt, shadowed shapes of the nearest rocks remained dimly visible from where I lay, reasonably comfortable, in the long heather, while the loch far below reflected dimly the greyness of the sky above. The prattling of a nearby burn, a truly Highland symphony, made music in my ears all that night, waxing and waning with the light breeze. Snipe (called in descriptive Gaelic 'the goat of the air') were bleating all night until the last two hours before dawn. With the last of the evening light a file of red deer fed in over a distant ridge top, silhouetted blackly to vanish as they came down the skyline. Sandpipers I heard crying from the far-off lochside; once too I heard the piping of oyster-catchers as they flew overhead. Less melodious was the bass resonance of a nearby colleague's snoring, audibly demonstrating his superior powers of slumber. Not a sound, however, from the vixen throughout that night.

I must have dropped off to sleep shortly before dawn, and awoke to find a colleague standing beside me about to awaken me. Time to go out while still cloaked by the semi-darkness and lie in wait. The air held a distinct dawn nip as I took up a position on a ridge some distance from the den. A ring-ousel, the mountain blackbird, fluted continuously as the world began to awake again, monotonous at last, from a rock face above my position, and when once or twice it interrupted its song to chatter harshly in alarm I scanned the dimly seen slope with renewed keenness, but saw nothing. As the light strengthened, grouse called intermittently and a curlew bubbled liquidly over the marshy flat fringing the loch below, while a colleague swore later he'd heard his first cuckoo of that year. By the time full daylight came we were all stiff with cold, the chilliness of an early morning breeze ill appreciated, and we were glad indeed to light a huge fire of dead heather stalks, to warm ourselves and make tea over. No results that night, but a store of memories.

Another night was, however, memorable for its success. A neighbouring colleague had been almost driven mad by persistent lamb killing and the consequent and natural complaints of the estate shepherds. At long last a den was found, the cubs dealt with and a message sent to ask me to wait out all night with him. About eight that evening the head of a fox was seen, skylined, peering down from a ridge above the den which was located in the middle of a huge face of riven rock. Taking my rifle I set out to stalk the fox, but when I arrived at the top it had vanished. I sat for a moment looking across a narrow flat to another ridge beyond, and was just about to move on when I saw a fox appear at the west end of the ridge and come unhurriedly along it, nose to the ground. I was caught out in full view and could only freeze where I was and hope for the best. Immobility saved the situation; the fox did not see me, and when almost opposite me stopped and began to nose at the moss. I steadied the rifle as best I could in my sitting posture, tried a shot—and missed. The fox turned and ran, but towards me, carrying an object in its jaws. When about eighty yards away it stopped, seeing me at last, and I shot at once and this time made no mistake. Clenched tightly in the jaws of the dead dog fox, when I went over, was a dead rabbit which he must have previously buried and had returned to carry in to the cubs. Later in the dark of that night as we lay among the rocks, in uneasy rest at the den, the vixen came working her way through the cairn so quietly that neither my two colleagues nor our terriers suspected her presence. I, however, as usual on these occasions, unable to sleep, saw her long-jawed silhouette outlined momentarily above the den; but as I reached for a weapon so did she melt into the darkness again. In the morning, however, after going out to the same ridge above the den, I saw her come along it towards me at about 5 a.m. This time I had my colleague's telescopic sight-mounted rifle, an advantage in poor light. It had, however, also a hair-trigger with which I was unfamiliar. I began to swing on the vixen, finger taking, as I thought, a preliminary feel of the trigger, and the rifle went off, the bullet hitting rock just above the vixen. She of course ran immediately, while I, calling myself all sorts of an adjectival ass, ran hard in the direction in which she had vanished, with the despairing hope of perhaps seeing her again. Incredibly, as I came in sight of the long nose of the ridge running west at an angle below me, there she was, stopped, about 150 yards away, looking back. No time for niceties of position; I dropped at once beside a handy rock, a quick shot and the vixen joined the dog fox in full expiation of their lamb-killing forays that year.

Remarkable too was the night when *two vixens* were shot coming in to a den which held only one litter of five cubs. One vixen was shot at last light, the other early next morning. One was an old vixen; the other, a younger vixen, was obviously the mother of the cubs. But why the two coming in to the same den? A tentative theory was that the older one had lost her cubs in the first round of the den's offensive and, maternal instincts unassuaged, had joined up with the younger vixen (perhaps the cub of a previous year) in rearing her cubs.

There was the unforgettable night too when, in a year when many lambs had been killed, we waited out all night at the discovered den. My colleague on that occasion shot the dog fox at 3.25 a.m. and I bagged the vixen, whom I had heard start skirling soon after the shot which had killed her mate, ten minutes later as she ran past me, still wailing. If ever I heard a note of absolute anguish it was in the keening of that vixen. I would swear to this day that she had seen her mate killed, knew that her cubs were lost to her and was leaving the scene of the disaster keening in irrepressible misery.

I have no doubt that foxes are capable of feeling emotions thus; the pity is that they are killers *par excellence* from which lambs, the young of roe and red deer and the eggs and young of all ground nesting birds can expect little mercy when hungry cubs have to be fed. And so in the dearth of any natural control man has to do it. It is all very well to say that, if left alone, fox numbers would find their natural level, dictated by availability of prey. Available prey in the Highlands at present, however, is primarily lamb, and one cannot ever imagine farmers countenancing foxes finding their natural level at the expense of their lambs.

Let there be no fears for the fox, however; it is in no danger of extermination. Constant warfare over the past hundred years in the Highlands is only just keeping their numbers to a reasonable figure. Many would feel it is too high a figure. And always it is the fittest and ablest which survive longest to ensure the survival of this resourceful species.

Spring into Summer

7

May

THE VALIANT VIXEN
THE RAIDERS

A RESOURCEFUL and beautiful predator, agile and quick-witted, the fox has fascinated its human pursuers for generations. In this I am no exception, and so when I had an opportunity in May a few years ago to piece together the fortunes of a fox family, partly from observation and partly from deduction and circumstantial evidence, I felt impelled to write of this. In order to preserve the continuity of the narrative I considered it better to put it down as if witnessed by some all-seeing, omniscient eye. This, however, is the only purely fanciful part; all the main facts are as noted in that year and are typical of what may happen in the annual fox versus humans warfare in the Highlands.

The grey light of the mid May dawning suited admirably the starkness of the dimly revealed Highland landscape. A bleak heather-clad ridge, its skyline indeterminate in the scuds of early morning mist which drifted along its top, dropped steeply in a grey rock-scattered slope to a burn brawling far below.

To the vixen, however, picking her way daintily among the rocks, a cock grouse in her jaws, the scene was homely enough, terrain to which she was well used. Her dog fox had gone missing shortly after they had mated earlier in that year, victim of man's unrelenting warfare on the Highland fox, and she had run alone thereafter until she had selected a den in which her cubs had been born in early April.

She had then had a very lean time, lacking a mate to forage and bring food to her while most of her attentions had to be fixed on her new-born, helpless cubs. A night or two after their birth she was reluctantly forced out by the need to forage for herself. Only half a mile or so from her den the rasping of wire upon wire caused her to halt and to make a cautious investigation. Caught in a sheep fence by the overlapping of the two top wires, which had ensnared one hind leg, she found a young stag almost *in extremis*, but with still enough strength to heave spasmodically at the ensnaring fence. Creeping closer, the vixen was seen by the trapped stag, who tossed his head wildly, causing the whole fence to jangle and vibrate in his renewed struggles.

The vixen decided discretion to be the better part of valour and blunted her hunger that night on a couple of fat voles and a very young rabbit caught venturing too far from its hill burrow. By the following night the stag was dead, and for many nights thereafter the vixen fed right royally on this food supply, so providentially near at hand, without which she might never have reared her cubs.

These cubs were now almost six weeks old and she was able to devote a major part of her time to hunting for them, a difficult enough task without a mate to share the burden. The cock grouse, held retriever fashion in her jaws, was the sole result of a long night's foraging.

Almost at her den, a rocky cairn near the ridge top, the vixen suddenly bolted the last yard or so as a terrific spine-tingling swoosh seemed to split the air directly above her. Safely underground, she turned to snarl soundlessly at the golden eagle gaining height after its unsuccessful swoop. The eagle too had young to feed: only one, but one with a large appetite, and fox, to this particular eagle, was no novel prey.

Inside the den the vixen speedily forgot the eagle as five blue-grey furry bodies flung themselves forward to spit and quarrel over the grouse carcase. Leaving them to it she curled up in a corner, able to relax in underground security, and dozed, tired after weeks of single-handed ministering to her litter's needs. The cubs, their heads now beginning to assume the red fox colour in contrast to their still blue-grey body fur, continued to squabble until only a few feathers remained.

An hour or two later the eagle, now only a dot high above the ridge, saw the figure of the estate stalker toiling doll-like along the rock-strewn steepness of the glen face, with two terriers running, minute-seeming, in front of him.

Lacking the eagle's all-embracing viewpoint, the first inkling the vixen had was the faint, far-off rasp of tackets on rock, an alien sound which galvanized her into instant alertness. Almost at once the hurtling entry of two eager, fox-hungry terriers caused the cubs to scatter in lightning reflex action into the cracks and slits which seamed the cairn's inside, while the vixen bolted immediately. Activated by the overriding instinct to save her own skin or by the desire to decoy the terriers after her, who knows? In this case it worked both ways.

The hard-running stalker, whose heedless terriers had far outdistanced him on first getting the irresistible scent of the den, was just in time to try a despairing and unavailing long shot as the vixen ran, terriers hard behind her. Cursing, slipping and slithering as he strove, dangerously, to hurry along the steep face, the stalker floundered after the rapidly disappearing terriers, themselves outdistanced by the fleet vixen, who, confused momentarily by the closeness of shot and pursuit, went to ground again in a long scattered cairn half a mile around the hill's shoulder. There she crouched for a moment, recovering her wits, and releasing, willy-nilly, the strong taint of fox alarm, until the scrabble of blunt claws and hard breathing at the cairn's entrance apprised her of the terriers' arrival.

Snaking noiselessly along in the darkness of the cairn's labyrinthine passage-ways, she left by a crack so tortuous and narrow that no terrier could follow. When the stalker arrived in a lather of sweat minutes later she was already well away, while the terriers were still underground, engrossed in the strong scent she had left behind. The stalker remained hopefully on guard above while his terriers took their time underground, alternately coming to the surface and retiring underground again, striving to unravel the puzzle the vixen had left, eventually coming to the surface and making it plain they were no longer interested. Realizing the vixen had foiled his terriers and was no longer in the cairn, the stalker returned, cautiously this time, to the first cairn, and the sudden re-entry of the terriers caught one cub which had prematurely ventured into the main chamber. A nipped-off squeak told the man outside what had happened; but though he remained there for the rest of that cold, increasingly wet day his terriers had no further success, though they blunted claws and skinned muzzles in their efforts to penetrate into rocky cracks too narrow to admit of their entry. When he eventually left, the stalker, cold, tired and saturated, neglected, amazingly, to block the den's entrances, though he resolved to return next day with assistance to try again. He knew well enough that only an empty den would reward him, but hoped, optimistically, to pick up the trail of the decamped inmates and so find the fresh den.

From a commanding position in the long heather of the ridge top the vixen watched stalker and terriers go, and indeed stalked them for more than a mile on their homeward way before she satisfied herself that they were really departing. She then made her way back, fast, to her cubs, her long snout wrinkling in disgust at the doggy smell permeating her den. For a moment or so she sniffed gently at the dead cub and when she turned to leave the now dangerous den, followed by her four surviving cubs, she picked it up, as if unwilling to believe it dead, and carried it with her.

The small convoy journeyed far that night, through dripping heather, threading rock screes and somehow fording the burn in the glen's bottom *en route* to a new den. Small as they were the cubs were precociously agile, resting for brief intervals when their short legs tired. Well did the vixen know where she was leading them. Every likely cairn in the area had been prowled through at some time in her wanderings, and by the dreich wet dawning the tired cubs were again underground. But this time the family was split up, two cubs in a rocky slit of a cairn at the base of a low cliff, the other two in a scattered shallow cairn about four hundred yards farther on.

To this second cairn the vixen carried the dead cub, laying it down in an underground chamber. There it lay for a day or two until it became quite obvious to the mother that it would not stir again. She then carried it some fifty yards and buried it under a shallow covering of grey moss beneath an overhanging slab. Whether she consciously thought of it as a burial is doubtful. The instinct to dig, and to bury unwanted articles or prey, is very strong in the fox, and probably this alone motivated her action.

For three weeks the depleted family had peace. The stalker had indeed returned next day with a colleague and more terriers, but on a day of ceaseless rain and dangerously low mist, every burn swollen bank high, they had searched fruitlessly for hours before returning home drenched to the skin.

The cubs were much redder in coat now, growing apace and

spending more time, when things were quiet, outside the dens, so that tracks began to be worn, visible to the observant eye. Occasionally they wandered far enough afield to visit each other, but in the main they kept to where the vixen had put them.

A strange clanking sound alerted the cubs of the shallower cairn one day, causing them to scurry underground, the vixen being absent hunting. Peering out inquiringly they saw a lean dog fox dragging on one foreleg a small trap, its chain burnished bright by continual rubbing through heather and rock. Attracted by the increasingly odorous prey remains scattered around the den, he fed ravenously on some scraps of lamb and thereafter became an almost daily visitor, scavenging in his need for food. Since getting the trap on his foreleg earlier in the year he had existed precariously on carrion, deer and sheep carcases mainly; but carrion was now scarce, the seasonal mortality peak being past for that year.

The den's immunity was soon to end. In her necessity to feed four perpetually ravenous cubs the vixen had killed lambs. These lambs had been missed eventually and the hue and cry was up. After a day or two's searching the two estate stalkers, with two terriers, came upon the rock-slit den on a day in early June. Here underground, in the deepness of its slit, were most of the noisome prey remains, and the terriers, perhaps confused by the nauseating stench inside, failed to locate the cubs, deep in the narrow fastness. Returning above ground, however, they almost at once picked up scent which a visiting cub from the other den had left only that morning.

The stalkers followed the eagerly questing terriers and were led straight to the farther shallower cairn where the more obvious prey remains lying above ground convinced them that this was the main den.

Excited barkings from underground, where the terriers had almost at once found a cub holed up in a narrow crack, further convinced them. Throughout the remainder of that day, however, stalemate prevailed, the dogs barking at the cubs but unable to get at or to bolt them, and the vixen obviously away from home. One stalker therefore remained at the den while his colleague returned the long miles home for warm clothing and provisions so as to remain at the den all night. Poking around among the rocks of the cairn, shaking his head bitterly every time he found the remains of lamb or grouse, the remaining stalker unearthed the body of the cub the vixen had buried and 'wondered exceedingly'.

Long after the first stalker had gone, the trap-encumbered dog fox came scavenging towards the den and, detecting the remaining stalker at once, retired, giving the den a wide berth. But fate was well and truly against him, for he ran straight into the returning stalker, who, rifle slung on back, laden with food and clothing, was nearing the den. The stalker, hearing the telltale rattle of the dragging chain, was alerted, so that fox and man saw each other almost simultaneously. As the fox turned to run so did food and clothing strew the heather, while the stalker frantically clawed for his slung rifle. Hampered as he was the dog fox was too slow in his retreat and a merciful bullet ended weeks of misery.

About 10 p.m. the vixen, quite unsuspicious, approached the den. She was within rifle shot of one of the concealed stalkers, but to make quite sure he waited till she was closer. This decision undoubtedly saved the vixen, for as she momentarily vanished behind a hump of heather so did she get a taint of human on the fitful evening air, and at once taking cover in a convenient heather-concealed drain she did not reappear. She nevertheless stayed in the den's vicinity all night, and with the first glimmering of light worked closer and skirled once. As though waiting for this, both cubs shifted from inside the shallow cairn—an ill-fated move, for both were accounted for by the alerted watchers.

The vixen, as if realizing this, melted away before the strengthening daylight, accompanied by her two surviving cubs from the rock slit den. The stalkers left about 8 a.m., cold now, and tired, but reasonably certain that they had settled the issue. As indeed they had, as far as lamb-killing in that area went; for the vixen and her cubs, older now and capable of journeying farther, journeyed far out to the high ground before they stopped again, their home this time in the dry tunnels of a high peat hag. From now on their prey would be grouse and the occasional windfall of a new-born red deer calf, besides what other small prey in the shape of voles and fledgling birds the high ground could muster. The vixen was luckier than many of her species in that she had herself survived and saved two cubs, while the stalkers were happy in having stopped the depredations on their ground; honours had been shared.

A day or two later the eagle, seeing the fox carcases lying on the rocks of the cairn, carried them off one by one to its eyrie, where the hungry eaglet disdained them not. Little is wasted in the wilds at a time of year when many hungry youngsters are being reared by overworked parents.

Towards the end of May, with new growth at last becoming apparent on the hill, the red deer, coats of old winter hair shabby and bleached looking, will be noticed to be grazing almost without cessation throughout the day, whereas in full summer much of the day will be spent in lying up, resting.

With the new grass growing faster and sweeter on the low-ground arable while yet the hill grazing lags behind, and with the tender shoots of new corn beginning to appear, the stags, always bolder and inclined to forage farther afield than the hinds, may be tempted to raid arable ground adjacent to the hill. It is certainly not starvation which impels them thus; sheer greed for sweet pickings would seem to me a more tenable theory. What schoolboy after all would eat bread if he could, even at some risk, obtain cake? By the end of May, however, there must be a definite strain on a stag's constitution, imposed by the growth of his new antlers, and a deeper need than mere greed may impel him to seek more nourishing fare than is afforded by the largely acid hill grazing.

In late May one year some thirty stags discovered how sweet the young corn was in an outlying field here. On the boggy moorland to the west of this field stags had wintered for many years, finding quiet in the uninviting bogginess of the ground. The field was bordered on its moorland edge by a wide embankment of earth and stone, the moorland side sloping, but the field side having a steep drop of about four feet. A fence, at one time designed to keep the stags out, ran along the middle of this embankment, its rusted strands and leaning posts now incapable of keeping out anything. Across this embankment in fact the stags were coming nightly, lured in by the young corn. At first I tried scaring them out nightly, with much shouting, and saluting of the evening sky with gunshots. When, however, a strip of ground at the west end of the field began to assume the appearance of a well-kept lawn, albeit one indented with cloven hoof marks, it became imperative to take sterner action. Reluctantly I realized I would have to shoot one or two of the ringleaders of the marauding band.

One evening therefore towards the end of the month I took my rifle and went across to the moor, to have my plans ruined that night by a factor I had neglected to take into consideration. There were peewits (green plover) nesting on the fringes of the cornland, and by now some had young hatched. As I made my way cautiously into an enfilading position, on a shallow ridge where moor met field, these parent birds took exception to my crouching form and began wheeling over my position, blunt wings whick-whickering as they stunted above me, pee-weeping incessantly. By way of varying this a pair alighted in the field near by and kept up a continuous and nerve-wracking, long-drawn-out screechy note, the one answering the other. I lay, thus tormented, watching the stags about five hundred yards off, feeding slowly in along another ridge of the moor, at times silhouetted on its skyline, at others almost hidden against its slope. My forebodings about the incessant complaining of the peewits, loud in the evening's quiet, were finally realized when the leading stag at length lifted his head, stood stock still, black against the fading light, and looked and listened for ten long minutes. Would he become reassured or would he go? He decided to go, and turning at last started moving slowly out. It was no part of my plan to shoot a stag on the moor; I wanted a salutary lesson administered as they were actually in the act of entering the cornfield. I therefore merely followed the stags and chased them farther out, in the deepening gloaming, though I know they would merely stay out until the cloak of complete darkness emboldened them to venture in again.

The following evening at about 9.30 I set out again for the scene of the damage. It was again a clear, windless, near frosty evening, the clouds clinging close to the hill tops to north and west in long, pink-tinged streamers, tinted by the setting sun. I had left it a little later in the hope that the peewits would be settled for the night. I was, however, most cautious, stalking along stealthily below the old dike on one edge of the field, as if the field was indeed already full of stags and not just a scattering of peewits. One or two of these birds took wing as I arrived at my vantage point, but luckily soon settled again. From where I now lay I could see the easily discernible track which the nightly forays of the raiders had left, winding in and out of the scattered shallow pools which dotted the boggy flat, and slanting up the slight incline of the embankment. About half a mile away I could see the marauders, between thirty and forty stags, some with growing, velvet-encased antlers already above ear level. I watched the spread-out feeding beasts for a few minutes until there came about an obvious change in their formation; they had started stringing out, the leading beasts coming towards me on their usual route in. Greed was obviously overcoming caution; they were not going to wait for the coming of complete darkness. There was going to be no difficulty in identifying the leading spirit either; he was a good ten yards ahead of his fellows.

Retiring to a better, more concealed position, I was then unable to see the incoming deer, only the place on the embankment where I had determined I would shoot the first raider. Minutes later, with mounting excitement tinged with regret, I heard the quick, plashing, eager-sounding footsteps of the leading stag as he crossed the water-logged flat bordering the embankment. In a moment he came into my view, a pause at the embankment's foot, then up it, catlike in his agility, a contemptuous scramble through the rusty fence, and then a last pause to survey the field ere he jumped down into it. Regretfully I squeezed the trigger and he descended into the corn for the last time, quite dead. His followers, still in single file some ten yards behind, stopped short on the flat. I wriggled higher up on my ridge, and as they turned and began to trot off I fired at the last one, the erstwhile second in command. Running a few yards, he dropped with a splash in one of the shallow pools. The survivors, even after this second shot, stopped about four hundred yards away on the face of the ridge opposite me, loath to leave the sweet pickings, jostling each other in irritation, a shapeless jumble of beasts in the poor light. Two reared up as I watched, boxing each other with flailing forelegs, easing their tension thus. I raised my rifle sights and, taking careful aim at a huge rock some twenty yards from them, I fired at it, the noise of the rifle and the crack of the bullet's impact on the rock thunderously loud in the already violated quiet of that lovely evening. The raiders ran again, but once again, incredibly, stopped. I advanced now, in full view, up the boggy flat, marvelling at their intransigence. They bunched, milling uncertainly, but stood their ground until I was within a hundred and fifty yards of them. Running again, they now jumped the march fence and were lost to view in moments. Watching a distant skyline which they would have to cross I saw them eventually file across it, convinced at last, I hoped, that the hill afforded safer if poorer grazing.

8

June

THE CALF MONTH

WITH the coming of June Highland red deer are looking at their worst, their coats a travesty of the thick, dark winter covering which has served them well, now bleached and faded, unkempt with tufts of loose hair, with here and there bits of woolly under fur in the process of being shed clinging among the surface hair. Not yet, this month, the sleek beauty of the red summer coat, thin and body-clinging without the under fur of winter, simply a rather pathetic shabbiness which seems to underline privations undergone in the past winter and spring.

I watched a group of deer lying relaxed, enjoying their cud, one sunny day in early June. Many of the hinds looked as if wearing Dundreary mutton-chop whiskers, the tops of their heads and all down the long nasal ridge clean cut and dark in new hair, but the angle of the lower jaw white-whiskered in bleached, yellow-white old hair. I was irresistibly reminded in one case of the popular conception of a weather-beaten, dark-featured old ex-Indian Army colonel lying back in a barber's chair, chin white-lathered below dark visage. To heighten the resemblance, tufty white 'eyebrows' of old hair lingered above the hind's eyes. A yearling in this party was particularly untidy as it grazed near by, trailing bleached streamer like-wisps of winter under fur as it walked.

While April is the 'fox' month, August the 'grouse' month and October the month of the red deer rut, June is the 'calf' month to the Highland stalker.

Picture a huge natural amphitheatre cradled among high hills and remote from human habitation, its lower slopes channelled by watercourses and ribbed by heathery ridges, cut up, jigsaw fashion, by peat hags. High above, very high above, golden head glinting as it catches the June sunlight, an eagle glides, wings set, across the wide coire and out of sight. A magnificent spectacle to a human watcher, but one to inspire instinctive fear in the hearts of the very young, dappled, red deer calves, curled up motionless, hidden here and there in the coire's burns and hags, as their mothers left them when leaving for the higher ground early that morning. A typical June scene, repeated all over the Scottish Highlands as the annual calving season of the red deer gets under way.

Charmingly dappled and daintily formed, a red deer calf is one of nature's masterpieces. June, their birth month, the start of life for the species, is a month well judged to give the young calves a good start to life before the privations of winter arrive.

After a gestation period of eight months, then, from the rutting season of the previous October, red deer hinds give birth to their calves at intervals throughout June. A very few early calves may be born in the last few days of May, and an increasing number in the first week in June; but the middle two weeks in June are those in which most of the calves will arrive, with the peak period probably around 15 June. Red deer hinds do not all come into season together in October, but at intervals throughout the first three weeks mainly, hence the spread of consequent births in the following June.

A tailing-off in the calving will occur throughout the last week in June and into July. Even in August an occasional very late calf may be born. I have a reliable eye-witness account of a hind calving in late August, but such late calves seldom survive the rigours of winter. I have also two separate (and reliable) accounts of a stag mating with a hind in February and March respectively. These very occasional late matings, if conception did ensue, would of course result in abnormally late calves. A hind is on season for about twenty-four hours, and if she is not mated then will again come in season approximately three weeks later, repeating this perhaps four or five times if for some reason she is not mated. This must be rare, but it does make possible the occasional very late mating as witnessed by my informants. I have never heard of an authentic case of a red deer hind giving birth to twins, one calf being usual; but I do know of a stalker who once, and once only in forty years' experience, found twin foeti in a hind he had shot in the winter season. Whether she would have delivered both alive is another matter.

Most of the calving occurs, I believe, on what I will call 'the middle ground', that is in the coires at around the two thousand feet level, neither the low ground of the winter grazing nor the three thousand feet or so level of the summer grazing grounds. These coires afford more shelter, cut up as most of them are by watercourses and networks of peat hags, than the colder heights, and more immunity from disturbance than the more accessible lower ground, and are therefore more suitable for the first period of the calves' lives, the vulnerable day or two when they are obliged to be more or less static in the one area.

In June, with fresh grazing coming on the 'clean' ground of the tops, and with flies and heat becoming troublesome on the lower levels, most of the red deer will want to be high during the day. Any single hind seen *during the day* in the virtually deserted coires must lead to the supposition that she has a calf lying near by, or is soon actually about to calve. This is a good guide if one wishes to watch or find red deer calves. Paradoxical as it may sound, the best way to find red deer calves is *not* to look for *them*, but to watch likely single hinds, particularly towards evening, in the known calving coires.

Giving birth to a calf, a purely natural function, entails little difficulty for a red deer hind, though it does entail a certain amount of discomfort, if not actual suffering. Some hinds may leave their companions and give birth while alone, but I have seen hinds give birth while others grazed or lay near by, evincing no interest in the travail of the calving hind.

I watched one hind begin her birth pains at nine o'clock one June morning. She had been lying with three others, but left them, presumably as she became restless with the onset of her labour. Around the wide green arc of a high coire she walked, at times kicking out spasmodically with a hind leg, or nudging around suddenly with questing muzzle at her haunch as if trying to locate the reason for her discomfort, until she had left her companions a good half-mile away. Periodically while on this perambulation she lay down but, unable to settle, seldom for long. When near one shoulder of the coire's bowl she lay for a longer period, and I thought that there she would give birth; but no, she rose and began to retrace her steps, almost on the line she'd come on. This time her progress was altered slightly; at intervals she seemed literally stung into running a few yards, interspersing her progress as before, however, with periodic rests. Eventually she rejoined her former companions and lay down below them. Shortly after lying down she elevated head and neck, nose pointed in the air, and gave vent to a single mournful-sounding, bovine-like bellow, a sound which red deer hinds give vent to *only* at this their calving time and which *can* almost be confused with the rutting roar of a juvenile stag. The *exact* significance of the bellow I do not know. At one time I thought it might signify pre-natal pain, but I have since seen and heard hinds use it while returning to rejoin *recently born* calves. By the time a calf is a day or two old the mothers seem to revert to the normal muted, lowing call when they come in at evenings to rejoin their calves, a low-pitched call which is in no sense a bellow.

The calving hind now lay full length on one side and began to rub head and neck on the grass as if wallowing, now and again lunging out with a hind leg. To all this her cud-chewing companions paid no heed whatever, About 10.30 a.m. she jumped suddenly to her feet, and as she stood broadside I could see about six inches of two black-looking spindly legs with white tipped hooves protruding from her. Again she lay down, on her opposite side, in very obvious discomfort now, and again she bellowed on an indescribably mournful note, full of a desolate yearning. When she once more jumped to her feet I could see that she was a little further on and wondered at the time the birth was taking. A third time she lay, this time immediately at full stretch, scrubbing head and neck vigorously into the short grass. Came a violent convulsion, and simultaneously she fairly sprang to her feet and the calf dropped glistening from her. It was 11.17 a.m., two hours and seventeen minutes from when I'd first seen her.

The mother began immediately to clean the calf. I could see her tearing away at the membraneous envelope which partially encased the calf, and as she lifted her head a long white streamer depended from her jaws. Some of this she swallowed and some discarded. The nearby hinds, hitherto so disinterested, now evinced some curiosity, and first one then another came across, head outstretched on long inquisitive neck, to sniff at the new-born calf, to be rebuffed in no uncertain manner by the mother, who wanted no visitors at this stage.

Almost at once I could see the dark head of the calf bobbing about uncontrollably at first but growing visibly stronger. By 11.25 a.m. it was struggling to rise, and two minutes later it did get to its feet, but immediately fell again. Guided by pure instinct, however, the calf kept struggling to rise and get under the hind to suck, while the mother, unbelievably clumsy to my eyes, kept knocking it over as she shifted restlessly herself. By 11.45 a.m. the calf, legs still very unsteady, found the udder after one or two abortive questings below the hind's forequarters, and began to suck only to collapse at once, as if the shock of nourishment was too much for it. A moment or two later, however, and the calf was really feeding, a first feed which lasted six minutes, after which it wandered away a little and lay down. I then arose from my place of concealment and walked across, seeing first the unaccompanied hinds run, and then, after some natural reluctance, the mother, while the calf lay still. A female calf, it weighed at one hour old 12 lb., slightly below average, which is around 14 lb. for a new-born red deer calf. Her coat was almost dried out, but she still looked very dark brown, with rows of creamy-white dapples and her hooves irregularly edged, white and soft. They would become clean-edged, harden up and darken in colour as the calf used her legs; by the end of the first day most of the whiteness would be gone.

A new-born calf has no inborn fear of man and will not attempt to take flight, its overriding instinct being to lie still whenever anything strange moves in its vicinity. In thick heather or very broken ground this is fine, but there are many situations on the hill where a calf, lie it ever so still, sticks out like a sore thumb, such as in short heather, short green grass or the black of a wide peat hag. In all these situations I have found calves, trusting, rather pathetically, in absolute immobility, but pitifully obvious nevertheless. It is not in fact, as one may often be led to believe, their actual camouflage which makes it difficult to find red deer calves on the open hill. It is the huge area of broken ground in which they may be. Once one can pinpoint reasonably accurately the whereabouts of a deer calf there should be little difficulty in finding it, unless of course in very thick ground cover. I have pinpointed calves from some two miles away with my stalking glass and found them, though this takes some experience.

It is in the early period of their lives (the first twenty-four hours in particular) that calves are most vulnerable to predators—chiefly eagle or fox, both of which will kill deer calves on occasion. The mother, however, is seldom very far away at this period, unless alarmed by human interference, and a single high-pitched bleat from her calf will suffice to bring her running to it. I know of several instances of the collies of shepherds, which had chanced upon a calf while gathering sheep, being routed by hinds coming to the rescue of their calves. While ear-tagging red deer calves, I have known a squealing calf bring in as many as seven hinds from out of nowhere, all obviously with calves hidden in the area and all wondering if the squealer was theirs.

What I would regard as the classic *eye-witness* account of a red deer hind defending her calf from an eagle is given by H. H. Crealock in his *Deerstalking* (1880), in which the hind, rising up on her hind legs and striking out at the attacking eagle, ultimately routs it, though obviously near to breaking point with panic. Another eye-witness account related to me was that of a hind which had arrived back too late to save her calf from death, but had routed *two* eagles from its body. The stalker, who told me had been attracted by the noise of the hind's bawling, was spying from a distance and saw the whole pitiful scene. The hind, he averred, frothing at the mouth, was standing near her calf while near by, on a huge rock, sullenly sat a large eagle. Some distance away on the other side of the calf's body sat another thwarted eagle. The mother had obviously utterly exhausted herself in dashing at each eagle in turn as they made individual attempts to get back to their prey, in her anguish probably rushing at them perpetually, routing them temporarily, keeping them away from her dead calf.

On the whole, death from predators and those natural fatalities in the shape of still-born calves or death shortly after birth have, in their infrequency, little impact on the birth rate. A hind may 'mourn' a dead calf for a day or two, remaining in the area and returning periodically to her dead calf, apparently unable to accept it as dead, and in this situation also gives vent to the now appropriate mournful bellow. I have witnessed this in the case of a calf killed by an eagle (from which it was apparently routed by the hind before it had a meal) and in the case of natural death, a heart-tugging spectacle.

Fatality to the hind herself in giving birth is, in my experience, very rare. I have only found three hinds dead, in calving, in over sixteen years of watching deer, though I believe three such cases were once found in a *single* season on Rhum. Even this, out of some hundreds of hinds, is not excessive.

By the time the calf is twelve to twenty-four hours old, fear of human presence seems somehow to have become acquired, and while many calves of this age will still trust to lying still others will attempt to run, emitting a shrill squeal as they rise, and amazingly fleet of foot they are too. The trepidation of those who lie still is very evident in the heaving of their chests and the flaring of their nostrils, so that one feels almost ashamed of being human and arousing such evident fear.

It should be emphasized here that unless one has a definite purpose it is better not to handle calves or linger near them, appealing as they are.

When a hind is surprised with a very young calf, she shows a very natural reluctance to leave, and when she does the calf will usually follow for a few yards, depending on its age, only to fold up and lie where it drops as it finds the pace too hot. A very young calf in such a situation selects no hiding-place but simply drops in the tracks of the hind; an older calf, however, quickly acquires a sense of cover, and these calves will be seen to run off at a tangent, into available cover of long rushes, the course of a burn or a peat hag, before dropping. This is as good a guide as any to a calf's age and to whether it is likely to rise and run as one approaches. In the first instance it will not attempt to run, in the second it often will. When a calf is more than a day old the hind is liable to go farther from where she has left it lying for the day in the calving coire, and, certainly in hot weather, may go right out to the tops, returning in late afternoon. Then, in an otherwise empty coire, single hinds will be seen dotted here and there among its hags, *seemingly* engrossed only in their grazing. This grazing may last from one to two hours and then quite suddenly the pattern changes: the hind ceases grazing, lifts her head, looks all around and then takes a few eager steps forwards. She may stop again, almost as if sheepishly recollecting herself, and pluck another few mouthfuls, but her eagerness cannot be restrained for long and in a straightforward manner, which once seen is unmistakable, she will make a bee-line for where her calf is lying hidden, perhaps two hundred to four hundred yards away. A low-pitched call as she draws near and as if by magic her calf materializes from empty space and runs to meet her. If one lies concealed, with a stalking glass, on the opposite slope to a known

The vixen suckles her cubs, about one month old. The single cub (*right*), about five weeks old, shows the blunt head and dark fur of the young.

Head up, sniffing, the typical alert attitude of the hill fox.

Jumping a fence for the sake of the sweeter pickings inside its enclosure —a raiding stag.

Red deer stags sparring, with antlers in velvet.

Red deer stag in velvet in mid June, at the time when the red deer hinds are giving birth to their calves.

A hind bellowing the *calving bellow*, a sound heard only at the calving season.

The red deer hind sniffing at her hour-old calf . . .

'By 11.45, the calf, legs still very unsteady, found the udder.'

A few hours old. 'Charmingly dappled and daintily formed, a red deer calf is one of Nature's masterpieces.'

XXI A red deer hind grazing in the lushness of summer growth. When seen alone like this, in June, she is likely to have a calf near by; as indeed there was.

The golden eagle, the largest native British bird of prey, with a wing-span of up to 7 ft and weight up to 11 lb.

Only the head is golden: the rest of the plumage is rather drab.

The fully fledged young eagle in late July; hind legs of a rabbit lie beside it in the eyrie.

Below, left. A young peregrine falcon, fully fledged at six weeks old. Its immature brownish plumage is streaked downwards, whereas the adult plumage is barred across. The male peregrine is 15 in. in length. ***Right***. An almost fledged buzzard with four moles, prey brought in by its parents.

The common buzzard usually lays its eggs about mid April in the Highlands. It is probably the most common of the larger birds of prey there at present.

Much smaller than the eagles, the length of a male buzzard is 21 in. against 32 in. of the male eagle.

XXIII

A sika calf of about one day old which weighed only 4¼ lb.

A red deer hind cooling off in a hill burn in late summer heat.

Head up in alarm, a mature hind hearing a parasitic fly buzzing near by. The attacks of these flies ultimately drive the deer to the tops in July.

'Advancing into the river . . . the young leading hind drank from the black waters.'

calving coire, this most touching scene will be repeated time after time in the early evenings of mid June. The hind will stay in the vicinity of the calf all night, leaving it again in the early hours of the next morning as she joins the majority of the deer on higher ground.

There seems to be no hard and fast rule as to the age at which the calf begins following the mother always, at least until it is a fortnight old, when it should certainly be doing so. The exact age probably depends on the actual calf, and whereas one calf at three days old may constantly follow the mother thenceforth, others have been found lying alone at a week old, and I know of one case of a calf of a fortnight old found lying alone with the mother absent. Ages of these calves were known *exactly* as they had been previously ear-tagged, a practice whereby a plastic ear-tag is affixed to the ears with date and locality embossed on it. By so doing it is hoped to augment, in due course, our still rather scanty knowledge of red deer habits and longevity.

From a fortnight old, then, and probably in most cases from a week old, red deer calves will, with their mothers, join the herd of which she is a member, and from mid July onwards their dappled miniature forms will add appeal to the deer herds. At this age too, though still dependent on the mothers' milk, they will start doing a little grazing for themselves. The hind's lactation period is a long one; I have seen calves still sucking in the following February, but I have also known milk hinds to be dry in December. As the calves grow older so do they become more independent, and one may see a small group some distance from a herd, playing or lying together, a bonny sight to watch.

They will retain their dappled coats for three months, the dapples gradually fading until this juvenile coat is replaced by their first winter coat, shortly before the onset of the rutting season in the autumn.

9

July

MONTH OF THE YOUNGSTERS

IF ONE had to choose the most suitable month of the year for teaching young and inexperienced raptorial birds to catch the prey on which their very existence depends, then I would plump for July, a month full of the flutterings and rustlings of young life. In the matter-of-fact world of Nature the inexpertness of the recently fledged predator is in some degree balanced by the inexperience of the many furred and feathered youngsters of July. This, I am sure, is why Nature (or evolution, as you like) has arranged matters so that the young of all the various birds of prey I have watched in the Highlands, from small sparrowhawk to huge golden eagle, leave the nest in July. This is the more remarkable when one considers the widely differing incubation and fledging periods of these birds of prey. To compensate for this the golden eagle, with a lengthy fledging period for its young of ten to twelve weeks, lays her eggs in late March; buzzards and peregrines, with fledging periods of about eight and six weeks respectively, lay about mid April; and kestrels and sparrowhawks, with little more than a month's fledging time, lay their eggs in May. The differing incubation periods (i.e. from laying to hatching), plus these differing fledging periods, bring the nest-leaving time of all to July.

Mind you, the fact that young of birds of prey leave the nest in July does not mean that they are immediately left to fend for themselves. For a varying period of time (in the case of the eagle, perhaps well into September) the parents will continue at least to part feed them, while presumably the youngsters gradually acquire, by example and instinct, the knowledge to fend for themselves. No doubt a percentage of those slow to learn or clumsy in application fail to survive: natural selection at work once again.

July to me was often the culmination of weeks of watching the eyrie of a golden eagle, watching and photographing the various stages, from down-clad, through 'piebald', to fully fledged, in the growth of the young, and noting the varied prey upon which they were fed. Intensely interesting July was too, for from its first week the young were usually fully fledged and active on the eyrie whereas, when younger, an eaglet is often distinctly lethargic and liable just to lie motionless on the eyrie while one was at it.

On many occasions I left home very early in the mornings to visit an eyrie before my day's work began, and while July lacks something of the early morning freshness of May or June, there were nevertheless many memorable moments in the quiet of the July mornings also. Such as a morning when the mirror-like calm of a hill loch faithfully reproduced every fold of its surrounding hills, the quietness broken, but certainly not shattered, by the melodic piping of an anxious sandpiper; the tweeting of innumerable meadow pipits and the petulant bleating of a young red deer calf, temporarily separated from its mother and, rather unwisely, letting the whole world know this. Or the sight of a roe doe, only two hundred yards from my house, quietly grazing on a green rush-grown patch below my path, a lovely picture, glowing rust red against bright green. A little farther along the path that same morning a tiny shrew came running down the middle of the path towards me, jerkily, as if unsure of its way, tail pink and naked looking behind its dark-furred body. It either did not see or disdained to take notice of me, and when I put a foot in front of it, it simply ran jerkily around it. I watched it pursue its erratic course until it went into the long grass bordering the path.

Another glimpse of roe, at four o'clock one morning, when a startled roe buck sprang out of the bracken below my path. I watched it sail over a clump of bracken, disappear into the bracken's depth, and appear again in another gazelle-like bound to vanish in the green depths once more. This time he did not reappear, but I heard him give vent to a series of querulous gruff barks for some seconds after he had disappeared. A little farther along another, much younger, buck ran, but without the beautifully agile bounding of the first. He simply ran, slightly crouched and furtive looking, through some thin bracken ahead, vanishing in silence over a ridge.

Red deer of course were often seen. The little picture which lingers longest is that of a hind, very early on a lowering grey morning, a small, dark, distant silhouette as she crossed a ridge against the lightening sky, to be followed by the miniature silhouette of her very young calf. Nothing much in that—yet the sight of the small dark figures, seeming alone in the world in that early morning quiet amid the gauntness of the brooding hills, touched me deeply and etched itself on my memory.

Scents too seemed stronger on those early morning jaunts: the almost incense-like smell of birches after heavy rain and most evocative and characteristic of those July mornings, the sharp scent of wild thyme rising to one's nostrils as one trod through and bruised its profusion.

Approaching an eyrie one morning I was treated, unwittingly I'm sure, to a display of aquiline flight which was worth seeing. I first saw the female eagle dive perpendicularly down past the eyrie cliff, wings half shut and held close to her body so that she reminded me, in silhouette, of the Boy Scout badge. She came out of her dive and soared out over the glen, gaps left by missing flight feathers evident in her wings. As I watched I became aware that the male eagle also was there, higher up. He too dived then, in emulation of his mate, but, pulling out, vanished behind a ridge. The female glided, wings motionless, across the wide glen, and, approaching the opposite face, soared upwards effortlessly, then executed a quick turn which brought her gliding parallel to the face, coupled with a sideslip to lose height. A slow half-flap of her wings kept her under way along the slope, to soar up and sideslip once more before alighting, instantly unseen, on the rocky face. Her flying was a joy to watch majesty was there, effortless power and grace, and a complete mastery of the air currents—in her element indeed.

It was that day also that I first heard an eagle 'yelping', a very apt description, for the sound, incongruous as it seems, was incredibly similar to the yelping of an excited terrier. The eagle is seldom vocal when adult, and on this and on other occasions I am sure the yelping was voiced in irritation at my presence near the eyrie. This was emphasized to me the second time I heard this yelping when the female not only voiced her displeasure but also demonstrated it in unmistakable fashion. Upon that occasion I had arrived at the eyrie early on a dull morning, with mist sweeping low along the ridge tops. The eaglet, obviously full fed, was lying on the eyrie with traces of blood fresh on its already impressive beak, while two pieces of bloodstained breast-bone, of grouse or ptarmigan, lay on the eyrie's outer edge. As I stood at this edge the female eagle, her missing flight feathers identifying her, flew very close above the eyrie, then turning with fairly rapid wing beats she returned, below the eyrie this time, along the glen and out of sight. This she repeated more than once, flying above and returning below the eyrie level while I wondered and yet enjoyed the close view of her hugeness. When she flew, close as she was, below the eyrie, I was looking directly down at her immense wing span (7 feet in the case of a female eagle), a unique view of an eagle, and could see a lateral light-coloured band along each wing and a lightish, buff-coloured wide band (not white, the mark of the immature eagle) at the base of her widespread tail. Once I heard her yelping, on an exceedingly querulous note, as she went out of sight; the eaglet also heard, cocking its head inquiringly. I was looking at the eaglet through my camera's viewfinder when the female arrived once more, this time her arrival heralded by a tremendous 'swoosh' of displaced air as she pulled out of a stoop just above me. I ducked instinctively and glanced upwards in time to see her turning away, so close that I could distinctly see her powerful yellow legs folded back along under her body. All the remainder of the hour I spent at the eyrie that day she kept up her 'patrolling'; twice more making me jump as the 'swoosh' of her stoop gave me first indication of her presence. The eaglet seemed emboldened by the female's continual patrolling. At any rate, no longer lethargic, it became distinctly aggressive, standing on the eyrie clapping its wings over its back, dust and feathers arising from the eyrie's tawdry surface in the breeze so caused, advancing a little towards me, thrusting out head and open beak at me, adopting a vulturine, threatening posture, the feathers of its mantle standing lanceolate, separate and erect, behind its outstretched head. Occasionally it struck out with a foot, but more often it clapped its wings and simultaneously advanced, once coming so near that it struck my shoulder with the edge of one wing, more by accident than design perhaps.

While the female eagle persisted in her tactics of swooping close to my head I had pondered (with some justification, I feel) on what

I'd read of eagles adopting these tactics in attempting to drive panic-stricken deer over the edge of a precipitous hill path. Looking, insecurely perched as I was on the eyrie's outer edge, straight down into the glen with its river winding far, far below, brought home with some force the feasibility of such tactics. I was, however, not panic-stricken, simply somewhat daunted, but stubborn enough to persevere in taking the photographs I wanted, believing, in any case, that the proximity of the rock overhang above me would keep the female from really pressing home her attack for fear of injury to her huge wings.

Upon my next visit this eagle again attacked me, yelping again in her displeasure and swooping even nearer, and this, mind you, when the eaglet was fully fledged and was within a few days of leaving the eyrie. This behaviour, however, is definitely not characteristic of eagles; in most cases the adults make themselves very scarce when a human intruder is at or near their eyrie. In ten years of eagle watching only two eagles have attacked me thus, two different birds in different years and in glens widely apart.

Prey at eyries over a ten-year period has been very varied, with grouse figuring most regularly because, although not present in large numbers, they are the natural prey most readily available in the area. Mountain hares were taken, but so rare are they hereabouts now (in the 1920's their number was legion) that this favourite prey seldom figured often in a season. Rabbits also featured regularly, until myxomatosis wiped out even the hill stocks of these. Vole and black water vole also appeared, though irregularly; small prey for a bird with a seven-foot wing span, but so also is ring ousel which I also saw as prey. Ptarmigan I saw only occasionally, and once I saw a large stoat lying dead on an eyrie beside its two eaglets. In practically every season fox turned up at an eyrie, sometimes adult fox at that, though whether picked up already dead as carrion or actually killed I do not know. In one year I saw three fox cubs on different days at one eyrie. Two were live captures taken, I believe, from a high-ground den which I subsequently found and had to deal with. The third cub was lifted, dead, from where it had been left on a rock above the den and afterwards recognized when I visited the eyrie. A colleague once witnessed wildcat kittens at an eyrie in this area. Red deer calves began to feature at most eyries after mid June (i.e. when the calving peak was on), but at the eyrie of one particular pair I never did see deer calves, though it was in the heart of a calving area. Nor did I see lamb remains often, only some half a dozen times in ten years, and in most cases it was undoubtedly carrion. Nevertheless a rogue pair of eagles will on occasion take live lambs, but this should not be taken as typical of the species as a whole, nor as reason for its condemnation.

Eagles are believed to mate for life, and are very long-lived birds if allowed to survive. An instance of the fidelity of an eagle to its distressed mate I heard of, at first hand, from a neighbouring colleague, whose word I would rely on absolutely. He had been out in May looking for fox dens accompanied by three terriers, one of which was a red Border terrier. Rounding a bend in a wild glen he spotted an eagle persistently diving at something and, hastening forward curiously, he saw that it was a fox. So engrossed were eagle and fox that he got close enough to attempt a shot at the fox, but without apparent success. No sooner did he shoot than his red terrier darted ahead after the disappearing fox. The eagle at once transferred its attention to the darting red form of the terrier and in a trice had it up in the air. My friend stood agape for seconds, then collecting his wits put a shot in the air *above* the eagle in an attempt to shock it into dropping his terrier. In this he succeeded, but his dog was unavoidably injured in the fall as well as nastily holed by the eagle's talons. Carried home, the dog, I am happy to say, completely recovered. This was not the end of the episode, however, for a few yards away, hitherto unseen in the peat hags, the stalker found the mate of the attacking eagle, grounded, and with a gin-type rabbit trap fixed on one leg. By it lay the remains of two lambs, obviously brought in by its mate. The fright engendered by the appearance of a human so near induced the grounded eagle to make a supreme effort, and it contrived to fly, trap dangling, while the astonished stalker watched. One can only hope it got rid of the trap, even at the expense of a talon or so. How faithful was its mate, in feeding it and in attempting to guard it from first the fox, whose intentions would hardly have been 'honourable', and then the terrier, associated undoubtedly, in its sudden appearance, with the fox in the eagle's mind.

If journeys to the eyries at times provided memorable experiences, the journeys back also did on occasion. Coming home one afternoon I found myself very close to a few hinds and very young calves. It was a sultry sort of day, and even at 2,500 feet altitude the deer were being pestered by flies, including, obviously, the nasal bot-fly

(*Cephenomyia auribarbis*), a fly with a particularly loathsome life-cycle, to my mind. This fly, on the wing in July, injects a drop of fluid containing live maggots inside the nostrils of the deer, which maggots journey up to the base of the nasal passages and thereafter live, and slowly grow, feeding on the mucus in the nasal passages, until the following March, when they may be sneezed or coughed out by the host, to pupate in the soil and begin the process over again in the following July. Of this fly the deer, the adult deer more particularly, are mortally afraid, so that one sees them suddenly throw their heads up, snort violently, rub nostrils frantically into the herbage and perhaps rise and stampede madly off for some hundred yards. The adult deer seem to be largely successful in this evasive action, and I believe this fly victimizes mainly the very young and quite inexperienced calves and, to a lesser extent, the yearlings, for it is in these age groups that I have found the heaviest infestations. In stalking closer to watch the deer I disturbed a black-face sheep that day, and she ran from me full tilt, almost into a young calf and its mother. The outraged hind let out twice with waspish strokes of a lashing foreleg, thereby completely demoralizing the already startled ewe and accelerating her flight downhill.

My nicest and closest glimpse of red deer came, however, one day after I had just finished lunching, on my way home from an eyrie, while I sat back, dozing after an early morning start, against the gnarled trunk of a big alder tree on the river's bank. Half awake as I was, I was suddenly amazed to have materialize before me the head and shoulders of a young hind, followed immediately by some half a dozen other hinds, from over a low ridge and only five or six yards from me. I was in full view, but motionless, reclining against my alder tree, while the fire over which I had so recently made tea still faintly smoked beside me. The leading hind stood momentarily, eloquent nostrils aquiver, before advancing to the river edge, followed by another hind. The rest stood desultorily plucking at grass outlined, from my low position, against the sky. Advancing into the river, the young leading hind drank while belly deep in the black waters, a lovely picture before my delighted eyes. She was still very ragged looking (early July) with tufts of old hair loose here and there in her coat. Of the others, a mature hind appeared almost to be dappled, lines of old bleached hair 'dapples' showing in her sleek new summer red, while a hind near to her was altogether clean and red but for a few shreds of old hair along her backbone. The heads of all the hinds, lacking the thick winter hair, looked long and naked, almost reptilian, a thick vein running prominent along the side of the head, from nose to corner of eye, a vein never seen in the winter coat. The deer were restless without seeming to know why, and inevitably after a few minutes I was spotted and away they all went.

A very pleasing sight in July is that of stags, new antlers well grown, in velvet still, lying or standing, skylined, on a ridge as they invariably are on a sultry day, taking advantage of whatever breeze there is. Antlers impressively thick in their velvet, coats red and sleek, they make a bonny picture. Once I saw a very young dappled calf lying among eleven stags. There was not a hind anywhere in view, and when the stags went away in alarm the calf went with them. Cause for speculation there, but I have no doubt that the calf had only temporarily joined up with the stags and would find its mother later, dependent on her milk as it still would be.

Summer into Autumn

10

August

MATING TIME OF THE ROE DEER

AT ONE time August meant in the Highlands, above all else, the month in which grouse shooting began on the twelfth, and this it probably still does in those eastern parts of the Highlands where grouse stocks are still good. In much of the western half of the Highlands, however, the progressive decline of grouse and mountain hare stocks has been a mystery still without adequate explanation. Here in the 1900's it was possible to shoot three hundred brace of grouse in a season; by the 1920's only a hundred and fifty brace, and even in 1939 a hundred brace were shot. Now, in the 1960's, it would scarcely be possible to shoot twenty brace. Hills in the west are wetter, with heather scarcer and sheep stocks higher perhaps than in the east. This I am sure has something to do with it.

August nowadays means to me, and to the increasing number of people interested in roe deer, the month when the roe rut is well under way and when in consequence, even in the lushness of summer foliage, fascinating glimpses can be had of these captivating small deer. *Why* the roe rut should be in late July and in August, while that of red deer and Japanese sika deer is in the autumn puzzles me. Undoubtedly the best month for the birth of the young of deer in the wild, in the Highlands especially, is June, and it is then of course that they do mostly arrive, even with roe hereabouts. I know that some roe fawns are born up here in May, but a considerable number are born in June. Thus roe deer, a *much* smaller species than red deer, have a much longer period between their mating time and the subsequent arrival of their young. A well-established phenomenon known as 'delayed implantation' explains this apparent longer gestation period in roe. Following mating in July or August, the fertilized seed lies more or less dormant until about the end of December, when implantation occurs and development really begins. Nature has thus ensured that the young of roe *are* born at the most favourable time of the year, as with red and sika, but would it not have been easier to have ensured that roe had their mating time and, linked with it as it is, their antler casting and renewal times, at approximately the same time as red deer and sika?

Perhaps we should be thankful, however; for if the roe rut and the red deer rut coincided one would not be able to enjoy the interest of both, whereas now this is possible. It is most unlikely, however, that this was taken into account when the universal scheme of things came into being, and so the puzzle remains.

In the weeks of high summer undergrowth immediately preceding the rut one sees little of roe deer except perhaps of a buck, inquisitive of some momentarily inexplicable movement in his 'territory', and coming pugnaciously to investigate, only to retire, barking bad-temperedly, when he finds out. Does are very secretive when they are about to give birth to their fawns, and the high cover then, of bracken in particular in the wooded glens here, lends itself to this secrecy. When lying in this kind of cover roe will sit tight while humans pass very near. I have watched, unsuspected, roe graze amid bracken and then lie down in it, to become unseen at once. I have walked thereafter in full view, *close* by the bracken where lay the concealed roe, and the roe has not moved, knowing full well that it was hidden, but of course not knowing that I had watched it lie down.

For the first month of their life fawns appear to be left alone a great deal by the doe, who returns from her foraging at intervals to feed them. In suckling a very young fawn a doe may lie down, a habit I have never noted in red deer, enabling the fawn to feed while also lying down. Unlike red deer, I have seldom seen a roe doe and her family lying *close* together, even directly after feeding, in Christmas card or calendar fashion. Very cover-conscious from a very early age, the young leave the doe and go to lie some yards away from where the doe may also lie in thick cover. Notwithstanding this *apparent* lack of affection the doe is a very tender mother who spends some time daily after suckling her young in licking and nibbling gently at them, while the young reciprocate in like manner. There is undoubtedly, as with red deer, an almost sensuous enjoyment derived mutually from this ritual, and at times a reflex action seems triggered off when a fawn is impelled to scratch vigorously with a hind leg, while the doe is nibbling briskly at a 'trigger spot' about its withers.

The fawns of roe at play are captivating in the extreme, skipping and bounding, rearing high on their hind legs, racing around in a wide circle, perhaps to run in and butt vigorously at the lowered head of the doe. The doe too may become infected with their *joie de vivre*, and she can cut as merry a caper as the dappled fawns, sharing their wild abandon, throwing matronly dignity to the winds.

By the rutting time the fawns will normally be following the does more, but during the actual period when she is actually in season and with a buck the fawns will again be left on their own. The young however, may not *necessarily* be very far from the preoccupied doe. An instance of this I witnessed one August evening while watching a buck with a doe which I knew to have twins. The buck lay, infinitely graceful, on a little ledge near a lone rowan tree which grew on the edge of a bracken filled hollow, while the doe lay nearer the tree. After a few moments the doe rose and became lost to sight beneath the tree's low-growing foliage. There suddenly appeared then, only about fifty yards from her, her hitherto unsuspected twins, grazing slowly towards her. The buck then rose and came over behind the tree, whereupon the doe emerged and walked down to below her fawns. They began to graze slowly down towards her again, but the buck, seeming jealous and as if waiting for this, ran forward between doe and fawns. At one moment he actually swept at a fawn with his antlers, at which the youngster shied quickly away. The fawns were both females. When I left in the fading light the doe was lying again, the buck keeping close to her and the fawns grazing at a discreet distance. It was interesting to note the variation in the coats of the 'family'. The buck, a youngish one, was beautifully red and sleek, the doe had a great deal of almost purplish-appearing old hair on each flank, while the fawns appeared yellowish-red with dapples quite indistinct now. Roe deer fawns seem much more independent of the mother from an early age (except for milk of course) than red deer calves; for whereas if one disturbs a red deer hind and calf, the calf, unless *very* young, usually follows the hind; in roe if one disturbs, say, a doe with twins, the doe will run in one direction and the twins will each run in separate directions.

In another year, on an August morning, I sat above a green gully and watched a doe emerge, closely followed by her buck, his head and neck extended, nose close to her rump in that almost lecherous-like pursuit characteristic of the rut. The doe wove her way across a bare slope in apparently aimless manner until she vanished in the birches on its lower edge. I continued to sit still and minutes later, out of the gully, as if jet-propelled, appeared her fawn, to race across the same bare slope but at right angles to the adult's course, disappearing in seconds. Just as I was about to move on the buck reappeared, alone, and came back into the gully, scenting along, nose to the ground, weaving here and there obviously, to my eyes, following the doe's back trail. He went into broken ground between heathery knolls, vanishing and reappearing, and eventually came out on top of one of the knolls, giving me a lovely view of his rust-red sleekness and his slim-looking, longish antlers. What had happened, I wondered. Was the doe trying to elude him, a little fed up with it all, so as to rejoin her fawn? It certainly appeared so, and she had apparently, at least temporarily, succeeded, or why had the buck been tracking along her back trail? The buck too vanished after giving me plenty of time to admire him, and it was as if there were no roe for miles around. Had I been moving through the hill no doubt this glimpse would have been denied to me; for deer are adept at picking up movement, which is why sitting *still* at some vantage point is often infinitely more rewarding than forever trying to see beyond the next ridge or that next clump of trees.

While most of my sights of the roe rut in the Highlands have been in August, the rut may be started farther south by mid July, and many roe experts believe that the yearling does in their first season trigger it off. On holiday in Dumfriesshire (a county rich in roe deer) I watched, at five o'clock one morning in mid July, a yearling doe feeding alone in a field bordered on its upper side by a high stone dike, topped by two wires, a formidable obstacle it seemed. On the upper hill side of this wall there appeared a buck, very obviously conscious of and seeking this doe. Back and forth he ranged, unseen at times behind the dike, trying to find a way through, to appear suddenly in mid air as he cleared the high dike in one spring-heeled bound. He made straight for the doe and began weaving about the field behind her, the doe acting coy, but obviously by no means averse to his attentions. I thought in fact that I might witness the pair actually mating, but in this I was disappointed, though at times while I watched that morning it seemed very near.

Roe bucks are said to be virtually monogamous, and contrasted with the red deer stag they certainly appear to be so. Yet one August morning I watched a buck grazing close to a yearling doe, apparently his consort. About two hundred yards below, from out of thick bracken, there appeared a mature doe whom I knew, and the buck at once forsook the yearling doe and, racing down the slope, began to pursue the mature doe. Weaving around in circles in the bracken they went, the doe keeping just ahead of the buck's outstretched muzzle, and after a few minutes disappeared from my view. Shortly afterwards the buck reappeared and began to graze alone, but whether satiated or disappointed I do not know; both does had gone and he didn't appear unduly worried. I suppose it is possible, indeed likely, that the younger doe had been in season first, and the buck had 'kept company' with her, but on the day in question when the mature doe appeared near at hand, and possibly in turn, in season, he had had no scruples in making advances.

This notwithstanding, I had a very striking instance of a buck's apparent fidelity to his doe on the evening of 12 August some years ago, near one of the scattered outposts of birch wood on the hill ground. Approaching a heather-grown hillock on the downhill fringe of the wood I unexpectedly found myself quite close to a young roe buck, on the other side of the hillock I was advancing to. He ran a few steps, naturally, on seeing me; there was no surprise at that. Surprise was to follow, however, for he stopped and then, *in the face of my continued advance*, he started hesitantly back *towards* me, the picture of alarm but refusing to flee. As I got over the hillock's summit I saw, a very few yards away from me, snug in the long heather, a doe lying with her back to me, barely more than her head and big ears visible. This then was why the buck was returning. Devotion to his mate? Or simply inflamed beyond reason by his mating ardour? One prefers to think the former, but it may be that the latter is nearer the mark. Whatever the motive, he was courting danger, which he would normally never have done and which quite certainly, I believe, no red deer stag would have done. As I looked at the doe lying so close to me she, belatedly realizing from the buck's attitude that all was not well, looked around and, seeing me, leapt up and ran. She made off straight away from me, but the buck, instead of joining her flight, gave me cause for greater astonishment by immediately heading her off, herding her back towards where she had been lying. She of course tried to break past him, only to have him, neck outstretched, race alongside and again, by crossing immediately in front of her, force her to turn. Time and again this happened while I stood watching, *in full view*. At last, by feinting one way (and how neatly it was done) and suddenly jinking the other way, she eluded the buck and raced out of sight over a nearby ridge, hotly pursued by the buck. I resumed my homeward way marvelling, and then suddenly the doe was back over the ridge, almost on top of me, but without seeing me. She was being herded back by the buck, and both stood just my side of the ridge, while I, now lying flat in the heather, watched. About five yards apart they stood, both very blown and distressed looking, mouths gaping open, the doe standing in a hunched-up, tuckered-out looking posture. They stood thus for a few minutes while I continued to watch, then the doe, recovered slightly, again attempted to get back over the ridge, but again the buck headed her off. Suddenly she shot away and, taking him by surprise, by sheer acceleration got over the ridge ahead of him. I waited for some time, wondering if he would be successful in again driving her back, but they did not reappear and I had to leave for home. I was already late, as so often happens when out on the hill, I'm afraid. I could only assume that the rutting fever so possessed the young buck that he wanted to confine the doe to that particular area even in the face of the apparent imminent danger of my obvious presence, while perhaps the yearling doe was instinctively trying to make for the lower ground area where she had been reared, an area which would presumably be in the 'possession' of her mother and her particular buck. A red deer stag in such a case, knowing of the proximity of another stag, would do his utmost to steer his decamping hinds away from that direction, though certainly not to the extent of driving them back straight at a human intruder. Perhaps the young buck had been similarly motivated.

11

September

STALKING IN THE TAWNY HILLS

BY EARLY September the warm tints of autumn are already appearing on the hills. Here and there patches of bracken show red among its green acres which cover all too much of the lower hill slopes, so that one is constantly spying at them in case it is the red coat of a roe deer. Among the often sombre crags of the hill glens one's eye is riveted by the almost unbelievable splash of vivid red which marks the leaf change in the rowan trees, earliest of all to don their autumnal mantle. The birch trees change by degrees, the high-ground ones assuming autumn colours while the low-ground trees are yet green. On the hills the slowly fading heather still tints the landscape a royal purple, while on the wetter flats the now coppery tints of the deer grass accentuate the warmth and richness of the autumn hills. Amidst these warm hues the deer are far less apparent than against the short-lived prevailing green of the hill in July and early August, their red-brown coats blending where formerly they had been contrasting with their background.

Red deer will be looking at their very best in September; the mature stags should have their new antlers clean of velvet now, and with red coats sleek and haunches rounded off with good summer feeding will be in prime condition for the coming rut. The hinds too will now be looking well, all of them, 'milk' and 'yeld', in summer red; but whereas the yeld hinds will have the rounded haunches of the stags, the milk hinds will instead have put much of their summer grazing into milk for their calves. These calves should now be acquiring their first winter coat, indeed any calf still markedly dappled in late September is a late born one, with consequently less chance of thriving.

Stalking of the stags will largely begin in September on most Highland estates, though some estates favoured with suitable ground to hold stags in August will start then. This seasonal stalking of the stags is looked upon traditionally as a sporting highlight of the year, and so indeed it is; there is also, however, the very real necessity of keeping the natural increase in stag numbers in check, exactly as with the hinds. I personally believe stalking, if done skilfully, to be as humane a way as could be devised for dealing with the annual natural increase in deer numbers. Practised skilfully, the quarry has no inkling that it is the object of pursuit, and in fact the least suspicion that a human is about will lead to that same quarry being no longer there when your firing point is reached. The end therefore should come without apprehension or suffering, more than we can say for the end of many humans. The unfortunate fact, however, is that stalking is not always done skilfully; stalkers, and perhaps particularly 'rifles', vary in skill, and it is here that distasteful episodes may creep in, in wounding instead of killing cleanly. It should be an understood law that *every* endeavour must be made to secure a wounded beast, unless it can be ascertained that it is only superficially hit. This means to me following across the march if necessary, though every effort should be made to avoid spoiling a day for your neighbours. I see little merit in only following a wounded animal just so far over the march and then leaving it and phoning next door in the evening to tell him the luckless deer is on their ground. This method involves, at the least, a lengthy period of suffering for the wounded animal, and much may happen to preclude further pursuit by next day. Far better to see the job, distasteful as it will be, finished personally.

The 'rifle's' day will be spoilt by the wounding of a beast, if he has any feelings at all, and he may well be left for hours while the stalker follows up the beast. The stalker's day will also be ruined, for to be a stalker *should* imply regard for his deer. He will hate to see a beast wounded and crippling away instead of cleanly shot; he will be full of overwhelming anxiety as he follows up in case he cannot end the wounded beast's suffering, and he will have much additional lung-searing work in following up and, when the beast is eventually secured, in getting it home. He will also be very conscious of the misery of his 'rifle', in ninety-nine cases out of a hundred, at making a bad shot and inflicting suffering on his quarry. Should the beast in question go into ground from where it is impossible to get it out, it goes without saying that it should still be shot; the criterion now is to end the suffering of a wounded beast.

Point No. 1 in stalking therefore should be: 'Be sure of the absolute accuracy of your weapon, and have sufficient practice with it to ensure perfect familiarity and knowledge of your capabilities with it.'

A rifle may be an old favourite; it may have been your father's, your aunt's, even your grandmother's. Herein lies a danger, for rifles, like ancestors, get worn out, but, unlike ancestors, their life can be renewed with every success with appropriate attention. Rebarrelling a rifle will ensure it new life and accuracy. Zeroing should always be a point to be meticulous about. There is no excuse for the plaints I have sometimes heard that 'she shoots high, or low, or to the right, or left'. Ensure also that your usual grain of bullet is always used, for this can also affect the zero of your rifle, a fact that is not always realized. Stalking has come a long way since the use of muzzle loaders, and black powder meant that wounding was often an expectation, and dogs a necessity in securing these beasts. One had only to read Scrope's descriptions of deer stalking in the early 1800's to realize this. With modern high-velocity, very accurate rifles equipped with telescopic sights there is little excuse for missing except that of human fallibility. There are people of course who, for one reason or another, never make good shots; hard as it may sound, I believe these people should realize that deer stalking is not for them. If they must stalk let it be with a camera and a telephoto lens.

It may sometimes happen that a stag is mortally hit yet not dead when one approaches it. It is then far better, more humane and much safer to administer the *coup de grâce* with a bullet in neck or head rather than to use the knife. I once had this lesson very forcefully driven home when a dying stag whipped round viciously with his antlers, gashing my hand, knocking my knife flying and nearly disembowelling me.

A ghillie once told me of how, when out with a stalker, they had lost a stag which had dropped to the shot but afterwards got up and made off. Next day he and the stalker went out, in different directions, to continue the search. Angus, the ghillie, was searching a steep-sided burn when he came upon the stag lying in a gravelly pool, still alive, only head and neck showing above the water, and nearly three-quarters of a mile from where he'd been shot. He looked very far through, and so Angus decided to pull him out unaided, instead of going for the stalker. The pool deepened very suddenly just beyond the stag, into a black, deep hole. Wading in and seizing the stag's antlers, Angus found that he'd caught a Tartar. 'The stag was so weak', said he, 'that he was apparently only flicking his head from side to side, yet I was being flung about as if I weighed nothing, and could get no secure foothold on the slippery stones of the burn.' He dared not let go now, for fear of losing balance and going into the deep water (he could not swim) or of getting an antler thrust as he floundered off balance. He began to bawl for help, and luckily, for the stalker was far out of earshot, two shepherds who were going to the hill heard him. Between them they dispatched the stag and rescued Angus, who thereafter never put much faith in 'grabbing the bull by the horns'.

Much of the charm of stalking lies in the sights seen while out on the hill. One can say that these sights can still be seen without going stalking. Humans being as they are, however, few people will walk often to the hill with its long miles and steep inclines, unless they have a strong motive, which stalking provides.

I once watched two young buzzards stunting, practising stooping and the lifting of 'prey' (tufts of grass and earth, in actual fact) from off the ground, dropping the small clods when high in the air and attempting to catch them ere they reached the ground. As a variation of this hunting practice one would mock attack the other, the attacked bird almost going over upside-down and thrusting its talons straight up at the attacker. On another occasion we had a young eagle alight near to us as we lay watching deer and begin 'cheeping' lustily, ridiculous sound from such a huge bird, soliciting food from a parent eagle flying high above. Eagles feed their young for many weeks after they leave the eyrie.

Skeins of geese we see every year, and occasionally whooper swans as they pass over in their autumn migrations. The deer are, surprisingly, sometimes liable to stampede if these large birds fly low near them. I believe they associate them with the eagle. A colleague once watched a skein of geese scatter in all directions when an eagle appeared above them. With geese flying 'every which way', geese and eagle went out of sight before he saw the outcome.

Ptarmigan on the high tops, lying until one is almost crawling among them, then rising in a snow flurry of dazzling white wings, are a sight I never tire of. A flock of golden plover, wheeling in unison with sharply angled wings; or, on the lower ground, an opulent blackcock, rising and flying powerfully overhead, lyre-shaped tail spread, all lend delight to one day or another.

A fox mousing, on tiptoe, back arched and ears pricked, rising in an arcing bound to pounce, scrabbling frantically as its quarry eludes it; a wildcat lying curled up asleep in the shelter of a burn down which one is creeping, or an otter, on 'migration' through the hill from one river system to another. Rare sights, but to be seen by the fortunate on occasion.

And always the deer: a hind 'slapping' a recalcitrant calf with an admonitory foreleg; two irritable hinds rising on their hind legs and flailing at each other with their forelegs; a yearling nipping in and drinking from one side of a hind while her calf drinks from the other. Or just deer, lying or feeding in the sun, or, the other extreme, wraith-like on a ridge ahead while the mist seems unable to make up its mind whether to reveal or conceal them. To those keen on deer in fact the mere watching of them never seems to pall.

One wonders at times, as one creeps to a certain knoll, how often previous stalkers had done likewise. I once found a huge brass ·500 cartridge case at the base of a small hummock where we were lying with our modern Holland & Holland ·244 rifle. This old case dated back to 1900 or so, when these black powder express rifles were in use here. 'When they were fired, a cloud of smoke surrounded "rifle" and stalker for moments afterwards.' Thus an old friend who had been a ghillie in those days.

Stalkers vary enormously of course; there are those who have the interests of their deer always at heart, a majority, I hope, but there are also those who can stalk well and shoot well, but to whom the deer are subordinated to the shot. To such a stalker in fact the thought of coming home without a beast is gall and wormwood, so that selection goes by the board and much harm can be done. Such a stalker is much less to be desired than one who is perhaps less skilful yet has more feeling for his deer, and is always careful in the selection of those to be shot. Better by far to come home without a beast than with the wrong one.

Characters there are and always will be. 'Mac', a stalker at one time on a neighbouring forest, once took a 'rifle' to a stag lying, admittedly, in poor light. The recumbent stag was nevertheless all too obvious to Mac, but was quite indistinguishable to the unfortunate 'rifle'. In fact after much peering and craning of head and neck, what time increasingly exasperated directions came from Mac, the 'rifle' said: 'That's not a stag, that's a stone.' Mac, exasperated beyond endurance, replied with the trenchant words: 'Hell, man, did you ever see a stone flap its lugs?'

Mac had a man out on another occasion who missed a shot at a stag, also lying down. The stag, in no whit put out by the shot, rose, shook himself, turned around unconcernedly and lay down again rump on this time. 'He's laughing at you,' solemnly asserted Mac.

Autocrats there are too, as the very able but short-tempered MacDonald who one day had the misfortune to have that type of impatient 'rifle' out with him who is forever getting one step ahead of the stalker. MacDonald spoke to him twice, asking him to keep in single file, and was waxing increasingly wroth as the 'rifle' persisted in getting ahead again. A Gaelic speaker, his English was apt to let him down when he got excited, and at last he spoke sharply to the 'rifle': 'Look here, sir,' he said, 'who is stalking here, you or I?' 'You, of course,' replied the 'rifle'. 'Well,' said MacDonald excitedly, 'come back into my heel here or we might as well go home now.' And back the 'rifle' came and eventually got his stag that day.

Stories of apparently invulnerable stags crop up every so often in stalking circles, stags which have thwarted pursuit year after year. I had a 'rifle' out here one season who had, while in his teens and still at school, stalked a 'hummel' stag for three successive seasons without bringing him to book. He always had to leave in mid season for school, and a few days after his return on that third year a wire came to the school from his father; 'Fearfully sorry, shot Hubert.' Hubert was the elusive hummel.

Here I had a stag which eluded various 'rifles' for three seasons also, a stag with a very bad but very distinctive head, a remarkably wide switch, except that he had a small fork on one top. He came to the same herd of hinds, in the same place in a particular coire every year, as a 'master' stag will always when once he has held hinds. The 'forked switch', as I came to call him, was first seen and stalked, unsuccessfully, in 1955 and stalked twice, and missed, in 1956. After the season's end in 1956 I went round the coires taking a census of my stags and saw the forked switch alone about his usual spot. He was on very bare ground, but I decided to try and stalk him, just to see how I'd get on. Very much to my surprise I got in without much trouble and eventually lay within *ten yards* of the 'invulnerable' stag. It was only then, at close range, that I saw that his right eye was closed, with a dark stain of moisture seeping down his head from it. This, then, was why I'd got in so easily: his blind side had been towards me. I lay and watched and photographed the stag for over an hour, and saw that he was also badly crippled on one foreleg; there must have been the very deuce of a scrap for him to have been blinded, and crippled thus also. Away across on the other side of the huge coire a big black stag, the victor presumably, seemed to have every hind in it. Blinded in one eye and crippled as he was the forked switch was roaring incessantly and at one time savagely forked up clods of peat and heather high into the air. 'Travelling' stags dotted the coire, but none dared contest the stag with the hinds, though an ugly, narrow ten-pointer came across to where the forked switch was roaring. The indomitable 'switch' advanced to meet him and they paced side by side like two bristling terriers until the ten-pointer backed down and made off again.

I looked for the forked switch again in 1957, curious to know whether his blinding had been temporary or permanent. He was in fact in with the coire's hinds long before the rut (a thing I have noticed more than once with stags suffering from some disability; they seem to shun the company of their contemporaries and arrive in early with the hinds), and I saw that he was indeed still blind in that right eye. He was stalked, and shot at, at thirty yards on 9 September 1957 and escaped with a graze on his back; he then left the coire and on 26 September was back, and again unsuccessfully stalked. He was by this time looking in thoroughly bad condition, and I decided at the season's end to go for him myself. I spied him back in his usual spot in the corie, and blind eye or no he had a herd of twenty beasts. A long, long, very flat crawl got me in eventually, but when I got my sights on him I had to lower them again. What if I missed also? Stifling this incipient 'stag fever' (my one and only attack of this) I sighted again and the travails of the blinded switch were over. He was indeed in poor shape, thin, harsh-coated, right eye completely withered, with the bullet graze of 9 September crusted over on his back. There was no doubt he was better dispatched, but with his passing a character had gone from the hill.

Much advice on stalking has been published since Scrope endeavoured to initiate his contemporaries in the early 1800's, through the medium of his classic work *Deerstalking*, but perhaps a few words to end this chapter will not come amiss. First, always first, and foremost, as already said, your rifle: use one of adequate calibre, ·240 upwards. Ensure its accuracy and your own proficiency with it: your target is not an inanimate thing of canvas and wood but a living, feeling animal.

Footwear, rubber-soled shoes or boots for me, as light as consistent with comfort, rubber-soled for silence; the 'clink' of the tacket-shod traditionalist over rocky ground has saved many a stag. Any alien metallic sound is fatal in fact, so avoid metal-shod walking-sticks too, they make a devil of a noise on hard ground.

Clothing, nothing better than tweeds, of a drab hue but not too dark in shade as dark clothes become almost black when wet, and black is about the worst possible colour on the hill. Avoid the modern synthetic materials; these otherwise excellent garments make a dreadful rustle when crawling. Avoid also like the plague rubber, or rubberized, over-trousers; these will broadcast your whereabouts at three hundred yards to the deer and will so soak you with perspiration as to defeat their ends. A good buy is an ex-army paratrooper's camouflaged smock; this is a good colour and being wind and *shower* proof will be a comfort on bad days. This *particular* type of smock does *not* rustle noisily. The possession of a 20 × telescope will very much enhance the enjoyment of a day's stalking; binoculars lack the magnification for long-range hill work, though ideal for woodland use.

Move *very* slowly when near deer; quick, jerky movements are those picked up by the deer. When at the firing point don't just concentrate on the stag and ignore the watchful eyes of the hinds while so doing; keep your movements to a minimum and then only infinitely slowly. Too many careful stalks are ruined at the last moment by careless, quick movements. When you do shoot, eject and reload *at once* as with a reflex action. If a beast is wounded and you are at all doubtful of your capabilities in dealing with it, pass the rifle *at once* to the stalker. *Seconds are vital in a case like this.* Lastly, be guided by the stalker as the man who should know his deer and how to manage them; walk behind him in going through the hill. This is *not* a position of inferiority, it is simply and sensibly to present the smallest point of view to any deer at a distance. And in these days of short staffs, do help in dragging and loading the stag you have enjoyed stalking. This stag may have been shot in a place from which the stalker knows it will be the devil and all to take out, just so as not to disappoint you.

A last thought. Deer-stalkers nowadays should also be deer managers, a full-time job in all its facets of census work, calf-tagging, herd movements and the seasonal take-off of stags and hinds. Ability to stalk and shoot only makes half a stalker, I believe, one who achieves his 'number' year after year perhaps, but if in doing so he is shooting the wrong type of beasts, decline in their quality is inevitable, even if so slow as to be almost imperceptible. Continuity in management is very important on a deer forest; it takes many years to *really* know your ground and the deer on it. Estates which, for one reason or another, are continually losing dissatisfied stalkers cannot hope to have their deer managed as they should be. This is a long-term (one is tempted to write, a lifetime's) work in which one season's incompetence can ruin many years' good work.

12

October

THE MONTH OF THE ROARING

AS THE colours of the hill change so do the habits of the deer. The lower hills and coires, those below about two thousand feet, almost bare of deer in July and August, become peopled again as the hinds gradually drift down in herds towards their ancestral rutting grounds, those areas in which the deer have mated ever since there were deer. Groups of stags, still out on the higher ground, may be seen lying together, not yet at enmity and, as yet, apart from the hind groups.

Towards the end of September an occasional stag, usually a big one in the prime of life, may be seen quietly feeding with a large herd of hinds on the high ground. The rut is starting, and in the first week of October it really gets under way. The stags, comparatively silent throughout the rest of the year, find their voices, and the hill resounds as stag roars defiance to stag. Every day as October progresses sees more stags coming in from their summer grounds, until one wonders where they all come from, and where they disappear to after the rut is over.

The bigger, older stags are the first to rut, and they, coming in early, take over a very large herd of hinds. This herd becomes progressively smaller as more stags come in and take their share of the hinds, the original stag, much as he would like to, being unable to hold such a large number. A big stag will hold thirty or so hinds against all comers, and will not tolerate a rival near them, being ousted only by a stronger adversary, often only after a stern battle. The rut is anything but an easy time for the stag. Possessed by a seemingly unappeasable fever, constantly on the move, scarcely eating or desiring to eat, his mating time is hardly idyllic. As for the hind, she is completely at the stag's mercy, changing ownership with the stag's changing fortunes. By the end of the rut she may have had many masters. In this she has no choice.

Signs of the deer are everywhere now. Deer use peat wallows at most times of the year, indeed peat seems to play an integral part in the life of Highland deer. I have seen new-born calves nuzzle at and actually eat peat, and calves of a few days old go through the motions of wallowing on a stretch of exposed peat. Wallows are used in March and April when the itch of warbles and loosening winter hair may impel the deer, and in any spell of hot weather thereafter to cool themselves; but at the rut wallowing increases markedly, to become part of the ritual of it. In the very late stages of the rut, in early November, I have seen the ice on a peaty pool broken where deer have wallowed. On every boggy flat, then, in the rut, one comes on wallows, often only small, narrow pools in the peat, shallow waters cloudy with disturbed peat, with edges of a semi-liquefied, porridge-like consistency. A scattering of deer hair fringes the edge, banks are scarred and trampled by sharp hooves, yet in parts smoothed out by the rubbing of the user's body. Stags and hinds both roll at this time, stirring and striking their wallows with a flailing forefoot to get them to the consistency they desire. A stag emerging dripping wet and shining black from the peat is a fearsome sight—coat matted, neck thick and swollen, eye wild and antlers threatening. Let him only suspect human presence, however, and, fearsome as he looks, he is off.

The very heather is redolent with the scent of the rutting stags, a heavy, rank odour which clings to his coat and is transferred to the heather where he has lain, or where he has been thrashing at clumps of rushes with his antlers, leaving a wide circular area broken and devastated. The low-lying branches of birch trees and willow bushes also suffer from his excess pugnacity, branches being left broken, dangling by fibrous shreds, or completely barked, the ground below and around trampled and cut up by the sharp cleats of the plunging stag as he punishes the unoffending branch.

Having acquired his hinds, a stag becomes demoniac in his unrest.

His fever and desires give him no peace. He races constantly from one fringe of his herd to the other, to drive out an encroaching stag, emitting a series of enraged *staccato* coughing grunts as he hotly pursues the interloper. Having chased him far enough, he stops to roar threateningly before turning to trot hurriedly back to his hinds, just in time to launch himself frenziedly at another would-be abductor who has sneaked in while he dealt with the first.

After dealing with this one too he rounds his herd restlessly; seeming unable to bear the sight of a hind at peace. Singling one out he pursues her through the herd, head thrown back, twisting and turning, weaving an erratic pattern among the other hinds, who are quick to side-step as he approaches. Tiring of this at length, he lies down and, in sheer exhaustion, dozes uneasily, jerking awake on hearing a far-away roar, roaring lazily in answer as he lies, to be galvanized into renewed furious activity as he spots a hind straying too far, or another 'poaching' stag with designs on his herd. Little wonder that he loses weight and condition fast. If killed now his stomach, normally full and rounded, would be found to be almost empty, its meagre contents a blackish, evil-smelling, semi-liquid mess. The balance of his whole system is upset.

Intriguing sights reward the watcher daily. A roaring stag, etched black against a distant skyline, bellowing defiance to the world at large and to any prospective rival in particular. A 'travelling' stag (one looking for hinds), nose to the ground, scenting along like a huge black questing hound, in the tracks of some hinds which had gone that way only a short time before. Another traveller sniffing at the bed which a hind had vacated some time before, lifting his head high on extended neck, wrinkling nose and mouth in a grotesque, expressive grimace. Or the close-up of a roaring stag on a grey day of fine rain showers, his roar terminating in an explosive guttural grunt which propels a small grey cloud of breath before it, visible momentarily against the dark hillside before being lost against the greyness of the sky. A hind stretching long neck and head placatingly along the ground, looking upwards submissively, 'under her eyelashes' as her 'lord and master' approaches and stands over her reclining form. This attitude of utter submissiveness is maintained until the restless stag moves away, or the nerve of the hind breaks, causing her to rise suddenly and take to flight, seeking refuge in weaving among the other hinds. There is no doubt that at rutting time hinds may sometimes be in mortal terror of the despotic stag; hinds, that is, which are not actually in season at the particular time.

A rather comical sight is that of a stag trying to head off the retreat of his hinds when they have sensed danger and he has not. The leading hind making off, he races round, amazingly fast over a short distance, and heads her back to the herd. Another hind meanwhile seizes the chance to break away, and he has it all to do again. The hinds in their uneasiness are determined; the leading hind again evades him and is again chased, and while this is going on, one, two, then a group split off and race in the direction in which they wish to go. The entire herd now follows their lead, despite the frantic rushing to and fro across their front of the stag, unable to make up his mind which hind, or hinds, to try and chase back and equally unable to stem the rout. Finding himself eventually at the rear of his decamping herd he turns and gallops after them, no doubt cursing the whims of all females, still completely unaware of the watcher chuckling at his discomfiture, whom his hinds *had* detected.

Throughout the rut there will be small groups of male youngsters, knobbers, and two- and three-year-old staggies, lingering at a safe distance on the fringes of the hinds held in thrall by the master stags. Occasionally two of these aspirants will engage in mock combat, a fencing match in which they learn the feints and thrusts for later life. These mock encounters have no air of menace about them, temporary advantage not being pressed home. One of the contestants may tire and drift away, his opponent, anxious for more 'sword-play', may follow him and give him a *gentle* prod in the rump which usually has the desired effect. A third and older stag in whom the yeast of the rut is beginning to work its venom may wander across, the click of antlers rousing his latent ire. Sensing the awakening menace in the older stag, the staggies, sensibly, will break off, to resume perhaps their juvenile fencing at a safer distance.

Roaring, magnificent music of the rut, the very essence of the wild hills and glens which it fills at this time, reaches a crescendo as evening falls. Stags which may have been dozing uneasily through the middle hours of the day, or at least resting their voices, are becoming active again and roar answers roar as the gloaming deepens. Of one such evening I noted; 'Roars of every description, and some which defied description, echoed on both sides of the long glen that evening. Raucous roars, tortured, anguished roars, roars with a note of

While very young, roe fawns may suck when lying down.

As they grow older, they will habitually suck while standing, as red deer always do.

Family of roe deer in autumn; a doe and her twins.

XXVII

A roe buck closely following his doe at the time of the roe rut, mid July–mid August.

Ptarmigan among the rocks of the high tops, with the lichened greys of which their plumage blends superlatively.

Roe twins lying asleep in the absence of the doe, perhaps in her preoccupation with the roe deer rut.

A red deer calf in early September, the juvenile spots almost gone at about three months old.

The red deer hinds are looking well by early autumn; a trio of them, alerted, view the recumbent photographer with suspicion.

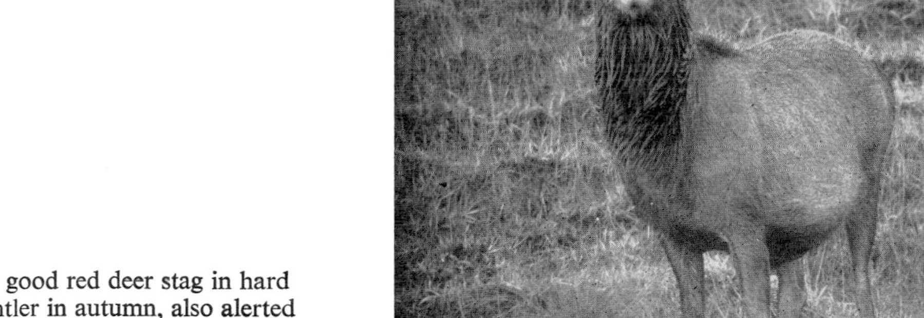

A good red deer stag in hard antler in autumn, also alerted by the photographer.

XXX A fine young eight-pointer and his hinds in October, the rutting or mating time of the red deer.

Young male red deer about fifteen months old, still *knobbers*, will be denied entry to the hind herds by the master stag at the time of the rut.

Two young stags sparring and learning to use their antlers. This game is not to be confused with the battles in real earnest of the October rutting time. Such was the scene below—a rutting battle which lasted fifteen minutes. The darker stag won.

XXXI

engage when he is conscious of being overmatched. In the more infrequent, serious fights, damage is often done, ears torn, forehead skin cut and bleeding, antlers broken, occasionally an eye blinded, while stab wounds in forequarters or haunches may cripple their recipient for some time. At times the battles end inconclusively. A few years ago I watched two huge stags who must have been fighting for a long time before I arrived. They were now plunging about on a very soft, boggy flat, head to head, actually down on their knees at times, so tired were they. Neither could master the other and, at last, by mutual consent, they separated and almost crawled away, mouths gaping and breath steaming, to take up stations on opposite sides of a very large herd of hinds which all the time had been grazing near by.

Regretfully eschewing some of the romanticism of preconceived ideas, it must be stressed that the real criterion in deciding the winner of most fights is nearly always bodily weight, and, as one seldom sees absolutely well-matched stags meet, most of the encounters are of short duration. It is downright chastening to reflect that a splendid royal or ten-pointer may have to give way to a heavier stag with perhaps an atrocious head, but so, unfortunately, it often is. Hummels, the hornless stags which, *if they live long enough*, usually achieve greater body weight than their antlered contemporaries, will triumph over these same contemporaries simply by virtue of this weight. A hummel is not invariably victor, however, as some seem to suppose. If he makes the mistake of taking on a heavier, antlered adversary he will lose the fight. I have seen a hummel, roaring loudly, trot over half a mile of hill to try his luck against an antlered stag who held hinds. The consequent struggle lasted only seconds, ending with the hummel leaving even faster than he'd arrived. This hummel was later shot and proved to weigh about 14½ stone and in age about five years; had he lived he would undoubtedly have achieved much greater weight and more success. On another occasion I watched a hummel being ousted from possession of hinds by an antlered stag of apparently similar weight, after, again, a very short tussle.

One does, however, occasionally have the luck to witness a really thrilling encounter between two evenly matched stags, and such a spectacle beats cock-fighting, prize-fighting or what have you, hollow. The sheer agility, absolute strength and tenacity of purpose exhibited are breathtaking, until eventually the loser concedes defeat.

Such a Homeric encounter I witnessed some years ago. It was towards October's end and I heard the antler-clashing as I walked towards a pass which split the hill ahead of me. A few hurried steps forward and there ahead of me on the ridge was a group of watching hinds and on the slope below them, head to head, two fairly evenly matched stags of around 15 stones, a yellowish stag and a darker, peat-stained one. The hinds saw me at once and made off, but the contestants noticed neither my advance nor their departure. Striving for ground advantage they plunged, heads always in contact, in wide circling, splay-legged sweeps, down the slope, ending up on the white-grass flat in the middle of the pass. There they fixed, still forehead to forehead, antlers branching on each side of their opponent's neck. Noses almost ploughing the ground with the steep angle of their necks, each striving with every ounce to push the other back. At times they were statuesque, momentarily in deadlock, then would come a lightning wheeling motion, heads still hard in contact and antlers becoming festooned with the high growing white grass. Now and again one stag would back away, literally dragging the other by the antlers so that one wondered how their necks could stand it. The occasional click and clash of their interwoven antlers was the only sound, not a grunt even from either stag, the very silence making the struggle the more elemental, and though I was now sitting in full view only twenty yards from them they took absolutely no notice of me. Now and then one stag seemed to get in his head at an angle which forced his opponent's head sideways, the attacker's brow-points dangerously near, or so it appeared, to the other's bent neck, until it seemed that the bent neck was forced at right angles to its body. When even their phenomenally strong neck muscles could stand no more of this the one at a disadvantage would wrest free, only to re-engage without pause.

It seemed, at one moment, that the fight was over when the dark stag actually drew blood from his opponent's neck with his brow points, but the yellow stag pulled clear and resumed battle at once. I watched enthralled as the struggle raged up and down the flat for some fifteen minutes, and then the tide began, slowly but perceptibly, to turn in favour of the darker stag. Soon he had the yellow stag going steadily back, step by grudging step, across the wide flat and then backing up the flanking slope. There, however, the wheeling tactics began again, their cat-like agility and utter silence almost uncanny as each strove to attain the upper ground of the slope.

Strangely enough the yellow stag attained this first and, in a flash, pushed his dark adversary, not steadily as before, but in one wild, headlong, plunging rush, right down the slope, the impetus carrying them again to the middle of the wide flat. It was his last supreme effort, however, perhaps born of despair, and also his undoing, for once they were back on the flat where ground advantage was cancelled out the darker stag first held him, then again began to push him steadily back, inexorably.

All at once it was over the yellow stag pulled free without warning and ran for dear life, leaping a drain in his flight. Without pause the victor was after him, breaking his former silence by grunting like one possessed, slashing meantime at the other's retreating hind quarters. Punishment he certainly inflicted, for quite a few swipes found their mark, so fast was he.

Fifty yards or so was sufficient for so hot a pace, and there the pursuing stag stopped, to roar vindictively, ribs heaving spasmodically and breath steaming. Roar followed roar, his mien so wild and threatening that I now rather regretted my exposed position. I was completely ignored, however. Turning and seeing no hinds left to reward his victory, he trotted away purposefully westwards towards where distant roaring proclaimed another stag, probably with hinds.

I have never seen stags actually charge each other, backing away as rams do and then a headlong rush at each other. A short, sharp run, yes, but more often stags walk for some yards on a roughly parallel course, seeming almost to be screwing up courage, before, it seems, simultaneously suddenly wheeling in one galvanic movement into each other and engaging. A preliminary fencing movement always follows as they strive to clear each other's branches so as to get their foreheads in contact, when the shoving match begins.

One sometimes hears stags credited with showing regard for a wounded stag. This I have never seen, quite the reverse in fact, though all my experience with this has occurred in the rut when the very worst in the stags' nature is uppermost. I have in fact watched a dying stag viciously attacked by another stag, with no hinds near at all, in a peat hag which the stricken beast had reached, and hidden in which his attacker had been. Even as the stag lay dead the other gored and plunged at it, almost in a frenzy, until he saw us approach, when he ran at once. It was then that we saw he was a crippled stag himself, and when he was eventually brought to book he proved to have an old healed injury which had stiffened and shortened one hind leg. I have known of two other cases of stags attacking dying stags thus; it appears to me that, at the rut, quick to sense any weakness in a stag which may heretofore have exerted superiority, another stag will at once take advantage and attack, chivalry having no place in the stags' rutting time.

As October ends the hill seems crawling with deer. Here and there small groups of hinds seem to have escaped from thraldom, but it is impossible to go far without sight or sound of a stag still herding hinds. By now, however, most of the older stags have had enough, and, their fever abated for another season, have left for their wintering grounds, leaving the field open to younger, less masterful stags. The hinds are in larger herds again and may have a number of young stags swashbuckling among them instead of one clear-cut master.

Lying concealed in the banks of a burn late one week-end a few years ago I watched such a herd. There were some hundred and fifty hinds and calves on a wide flat, and milling restlessly among them, roaring ceaselessly, were fourteen young stags, six-, eight- and nine-pointers. A little farther up on the flank of a hill an older stag, whose eleven-point head promised well, was having a hard time keeping his hinds from straying towards the big herd and its attendant escorts. With its continually shifting beasts and its babel of noise the scene reminded me of nothing more than a cattle market. The bellowing of some of the younger stags was almost bovine in character, but the mien of even the youngest was anything but bovine; all had shaggy necks and wild, staring eyes.

The hinds were scattered grazing about the flat, or lying chewing the cud. They were seldom vouchsafed peace to do either for long. One or other of the fevered youngsters weaving endlessly through the herd would approach, head thrown back, nostrils aflare and upper lip curled back, and the hind would hurriedly move off, pursued by the stag until he tired or a rival distracted his attention. Most of the hinds were very obviously fed up with the whole business, their own needs probably long since fulfilled; but they were, willy-nilly, chivvied here and there by first one stag then another, so that they had to keep watchful always as they hastily cropped at their grazing. Not so the stags. No one of them was strong enough to impose his will on the others or to escape with a 'cut' of hinds, and so there was stalemate in a constantly shifting kaleidoscope of movement. If one stag looked as if he was edging some hinds together he was soon thwarted by a jealous rival. Serious battle did not occur, but sparring and roaring went on continuously. Some stags engaged with a cursory clicking of antlers, others pranced at each other with heads lowered, only for one to bolt at the last moment. Others again advanced roaring, gauged each other up, and decided to confine themselves to vocal vituperation. The noise and activity were ceaseless it seemed.

As the gloaming came on the deer gradually worked in for their night's grazing on the lower ground, the instigators of the move being the hinds, followed by their caterwauling escorts. I left the scene as the darkness deepened, and the sound of the roaring followed me a long way through the hill. No doubt some of the more enterprising stags would manage to cut out a group of hinds in the night; certainly there was to be little rest for any of them.

By mid November the activity has markedly abated, although a minority of the hinds may yet come in season and roars may be heard throughout the month and, indeed, occasionally in December.

With autumn's splendour fast fading into winter's drabness the hinds start to wander out higher again, taking advantage of any open weather, until winter's storms drive them down once more.

The stag's wooing is over. There is little sentiment or romance in it, but there is grandeur, drama and spectacle which I, for one, never tire of watching.

I would end with a plea for our deer. Red deer have been subject to much abuse of late years. The sportsman who neglects to ensure that he can *usually* kill cleanly; the stalker who neglects to ensure that everything possible is done to secure a wounded beast; those who close their eyes to overcrowding in the hoary belief that the more deer the better; the irate farmer who blasts inaccurately, with whatever weapon comes to hand, at deer which are in his fields, or, worse still, sets illicit wire snares to the detriment of his walls or fences and the long suffering of the stag which goes away with a broken wire tight around him. The thoughtless boy who is out for a lark with a ·22 rifle; the progressive deprivation of the wintering grounds of the deer by forestry and other interests, without due provision. The poacher who battens on deer in the dark nights of winter and early spring when they are at their lowest ebb and the shooting contractor whose only idea of deer is simply so much per pound of venison, and whose methods may leave much to be desired. (I have heard of one such whose tracked vehicle ran down, in deep snow, the last survivor of a herd of seventy-eight deer, the driver having run out of ammunition.) The list is long and could be made much longer, and all on it are at fault in some measure.

Deer are attractive animals in their own right, worthy of being regarded as something more than sporting quarry, or venison at so much per pound, or vermin. For this they demand nothing but adequate shelter and subsistence in winter and spring, surely not an excessive demand.

Let all those directly concerned with deer reflect whether they are doing *all* they can in solving the various present-day problems arising with deer populations, or whether they are just drifting with the tide of bygone tradition, cosy in the realization that there will still be plenty of deer—at least in their lifetime.

THE AUTHOR

L. MacNally is forty-two years of age, Scottish born, and educated at Fort Augustus and Glasgow. He has lectured on wild life, and in 1967 won the principal award for colour photography of deer in the first annual photographic competition sponsored by the Midlands branch of the British Deer Society.

He contributes to *The Field*, *Shooting Times*, *Scottish Field* and *Scots Magazine.*

Jacket subject features Highland Stag photographed by the author.

Index

Roman numerals refer to plates